The Effect of Ocean Acidification on Skeletal Structures

The Effect of Ocean Acidification on Skeletal Structures

Editors

Hildegard Westphal
Justin Ries
Steve Doo

MDPI • Basel • Beijing • Wuhan • Barcelona • Belgrade • Manchester • Tokyo • Cluj • Tianjin

Editors

Hildegard Westphal
Leibniz Centre for Tropical
Marine Research (ZMT)
Bremen
Germany

Justin Ries
Northeastern University
Nahant, MA
USA

Steve Doo
Leibniz Centre for Tropical
Marine Research (ZMT)
Bremen
Germany

Editorial Office
MDPI
St. Alban-Anlage 66
4052 Basel, Switzerland

This is a reprint of articles from the Special Issue published online in the open access journal *Journal of Marine Science and Engineering* (ISSN 2077-1312) (available at: https://www.mdpi.com/journal/jmse/special_issues/F_Ocean_Acidification_on_Skeletal_Structures).

For citation purposes, cite each article independently as indicated on the article page online and as indicated below:

LastName, A.A.; LastName, B.B.; LastName, C.C. Article Title. *Journal Name* **Year**, *Volume Number*, Page Range.

ISBN 978-3-0365-8282-5 (Hbk)
ISBN 978-3-0365-8283-2 (PDF)

Cover image courtesy of Marleen Stuhr

Contents

Journal of Marine Science and Engineering

Editorial

The Effect of Ocean Acidification on Skeletal Structures

Hildegard Westphal [1,2,3,*], Justin B. Ries [4] and Steve S. Doo [1,3]

1 Leibniz Centre for Tropical Marine Research (ZMT), 28359 Bremen, Germany; steve.doo@kaust.edu.sa
2 Department of Geosciences, Bremen University, Fahrenheit Str.6, 28359 Bremen, Germany
3 Physical Science and Engineering Division, King Abdullah University of Science and Technology (KAUST), Thuwal, Makkah 23955-6900, Saudi Arabia
4 Department of Marine and Environmental Sciences, Marine Science Center, Northeastern University, Boston, MA 01908, USA; j.ries@northeastern.edu
* Correspondence: hildegard.westphal@kaust.edu.sa

Citation: Westphal, H.; Ries, J.B.; Doo, S.S. The Effect of Ocean Acidification on Skeletal Structures. *J. Mar. Sci. Eng.* **2022**, *10*, 786. https://doi.org/10.3390/jmse10060786

Received: 25 May 2022
Accepted: 1 June 2022
Published: 8 June 2022

Publisher's Note: MDPI stays neutral with regard to jurisdictional claims in published maps and institutional affiliations.

It is well known that the increasing partial pressure of atmospheric CO_2 (pCO_2) is reducing surface ocean pH, a process known as ocean acidification (OA). This results in a reduced saturation of the seawater with respect to the $CaCO_3$ polymorphs aragonite, high-Mg calcite, and low-Mg calcite that are involved in the biological formation of calcareous skeletons and shells. The effect of OA on calcium carbonate precipitation and subsequent dissolution in carbonate depositional systems, such as coral reefs, is a hotly debated topic. While early studies suggested that certain carbonate-secreting organism groups may be strongly affected by OA or even become extinct [1,2], others observed highly variable, species-specific responses to OA, whereby some taxa are negatively affected, some are positively affected, and others are unaffected [3–5].

Ries et al. (2009) [5] provided a systematic baseline study on the effect of OA on biological calcification by quantifying rates of calcification for 18 species of marine calcifiers cultured under controlled laboratory conditions that ranged from present-day to predicted year 2500 pCO_2 conditions, thus spanning seawater conditions that were highly supersaturated to undersaturated with respect to aragonite and calcite. These results illustrated that the biological calcification response to OA is variable and complex, depending on the species' ability to utilize elevated CO_2 directly via photosynthesis or indirectly via photosymbiosis, on the ability to regulate pH (and thus carbonate chemistry) at the physiological site of calcification [6], on the extent to which the shell or skeleton is protected from dissolution by external organic layers, and on the solubility of the calcium carbonate polymorph incorporated into the shell or skeleton (with aragonite and high-Mg calcite being more soluble than low-Mg calcite).

The relative tolerance of certain photosynthesizing calcifiers to increased CO_2 conditions was also demonstrated, for example, by the work of Vogel et al. (2015) [7] on the bryopsidalean calcareous green algae *Halimeda* growing on a volcanic CO_2 seep in Papua New Guinea at pH conditions down to <7. These *Halimeda* showed increased calcification in the light that compensated for severely reduced calcification in the dark, resulting in the same net calcification rate as specimens outside the influence of the seeps. At the same study site, interspecific differences in coral reef communities were observed in response to natural gradients in CO_2-induced acidification as one moves away from the seeps. Yet, despite the spatial variations in species composition of these coral reef communities, the extent of coral cover remained the same [3]. Similarly, Guillermic et al. (2021) [8] showed that increased CO_2 can lead to increased growth of some species of zooxanthellate corals when those corals are able to continue regulating pH and, thus, carbonate chemistry at their internal site of calcification. This underscores the variability of calcifiers' response to CO_2-induced OA, as well as the complexity of factors controlling these responses.

Although organismal responses to OA are well documented, community- and ecosystem-level impacts of have been less studied. Meta-analyses of organismal physiological responses to OA reveal large variability in how intra-organismal interactions (e.g., symbioses) may modulate responses to OA [9]. A growing body of research has also documented that OA alters competitive interactions [10] and favors the expansion of fleshy algae [11], which collectively can simplify ecosystems, reduce functional diversity, and impair ecosystem function.

The combined effects of OA and other environmental factors, such as ocean warming (OW) and food availability, on the capacity of marine calcifiers to precipitate and maintain shell or skeleton have also attracted increased attention over the past decade. Synergistic effects of acidification and warming range from amplifying negative effects to mitigating them [8,12,13]. These studies demonstrate that biomineralization is a process that is not merely controlled by external seawater conditions, but is also strongly influenced by the organism precipitating the skeleton or shell. These studies also emphasize the importance of incorporating other stressors in the marine environment that are likely to accompany OA over the coming centuries, including warming, pollution, eutrophication, and changes in salinity.

This Special Issue builds on this knowledge by presenting some current developments in research on the effects of OA on calcium carbonate producing organisms. The papers assembled here highlight novel methodologies and previously undescribed impacts of OA on calcifying organisms at functional scales ranging from the mineral, to the calcifying fluid, to the whole organism, and from temporal scales spanning the deep geological past to the future.

Although extensive laboratory- and field-based studies have shown that marine calcifiers exhibit highly variable and complex responses to OA, few controlled laboratory experiments have been conducted to evaluate the impacts of CO_2-induced acidification on freshwater calcifiers. Ninokawa and Ries (this issue) [14] address this knowledge gap with a freshwater acidification experiment showing that various common species of freshwater calcifiers, including a species of crayfish, mussel, and two species of clam, exhibit nearly as diverse a range of calcification responses to acidification as observed for marine calcifiers, with response patterns ranging from negative to parabolic. Their experiment also shows that freshwater acidification impairs the feeding rates of crayfish, which likely contributes to their negative calcification response to freshwater acidification by reducing the amount of energy available for their molt-mediated process of shell formation. Notably, similar trends have been observed for marine species of decapod crustacea, including for mud crabs whose rates of feeding [15] and calcification [16] were both impaired under acidified conditions.

Ries and co-authors (this issue) [17] report on a combined acidification and warming experiment on three Indo-pacific species of scleractinian zooxanthellate corals and one species of asymbiotic cold-water coral. They found that under normal temperatures, calcification rates of all three species of tropical corals increased with increasing pCO$_2$. However, when exposed to thermal stress that induced partial bleaching of the corals' symbionts, the tropical corals lost their ability to calcify more quickly under elevated pCO$_2$. Consistent with these observations, the cold-water species that lacks photosymbionts exhibited calcification rates that declined with increasing pCO$_2$. Notably, all four species of corals exhibited elevated pH at their site of calcification relative to seawater pH, with the offset increasing with pCO$_2$. However, for the one tropical species whose calcifying fluid pH was measured under both the control and elevated temperature treatments, thermal stress was shown to reduce the extent to which that coral species could elevated its calcifying fluid pH relative to seawater pH, a trend that was accompanied by both reduced symbiont abundance and decreased calcification rate. These results support the growing body of evidence that the temperature-induced loss of photosymbionts (i.e., bleaching) impairs the resilience of scleractinian zooxanthellate corals to the effects of CO_2-induced OA.

Eagle and coauthors (this issue) [18] report on controlled laboratory experiments investigating the complex physiological and calcification responses of four common Caribbean species of scleractinian zooxanthellate corals exposed to acidification and warming. They utilized the carbon and boron isotope composition of the skeletons of these corals to illustrate that dissolved inorganic carbon and coral energy reserves were lower in corals exposed to elevated temperature. They also report that these coral species' ability to regulate calcifying fluid pH in response to acidification was not impacted by thermal stress [8]—which contrasts the responses of Indo-Pacific pocilloporid corals reported in this issue. Collectively, these results suggest that the biological regulation of pH and other carbonate system parameters within the calcifying fluid of corals is energetically more expensive under pH and temperature stress. Although Eagle et al. [18] found that physicochemical control over coral calcifying fluid chemistry is linked to the physiological responses of both the coral host and its photosymbionts, they show that the extent and even direction of these relationships can vary amongst species.

Khalil and coauthors (this issue) [19] investigated the impact of acidification and warming on a species of starfish—an organism that, unlike zooxanthellate corals, cannot directly benefit from increased CO_2 via enhanced photosynthesis or photosymbiosis. They report that the Mg/Ca ratio of the calcite produced by the starfish increased with temperature, but showed no relationship to pCO_2. However, elevated pCO_2 led to potentially negative changes in test structure, with these effects exacerbated under elevated temperature.

Ocean acidification can not only slow the rate of calcium carbonate shell/skeletal formation, but also impair an organism's ability to maintain and repair their shell/skeleton after its initial formation. Bivalves, in particular, engage in extensive shell repair in order to mitigate the effects of boring parasites and predators, such as boring sponges and gastropods, respectively. Although the impact of ocean acidification on bivalve shell production has been extensively explored, the potential impacts of OA on mechanisms of bivalve shell repair are largely unknown. In this issue, George and coauthors [20] show that mussels are able to repair shell damage imposed by boring predators even under acidified conditions, underscoring the high priority that bivalves place on the repair of sublethal shell damage.

Scucchia and coauthors (this issue) [21] present a new artificial-intelligence-based method for assessing the effects of OA on the internal three-dimensional structure of coral skeletons. Numerous prior studies have shown that acidification can impact coral skeletal structure across a range of spatial scales [22–24]. Scucchia and coauthors build upon this work by using synchrotron phase contrast-enhanced microCT (PCE-CT) data combined with artificial intelligence methods (deep learning neural networks) to quantify how ocean acidification impacts the internal skeletal structure of coral recruits. They show, for example, that coral recruits exposed to ambient (i.e., non-acidified) pH conditions have less variability in internal skeletal structure amongst individuals than when they are exposed to acidified conditions.

Over geological timescales, OA has been shown to have impacted marine calcifiers in as wide a range of ways [25] as has been reported for modern-day calcifiers exposed to acidification in controlled laboratory experiments [5]. Pomar and coauthors (this issue) [26] reflect on calcareous biomineralization throughout the last 3.7 Ga of Earth history under highly variable atmospheric and seawater chemistry conditions. They critically evaluate the various hypotheses that have been proposed for the evolution of calcareous biomineralization, including for protection from predation, for structural support, for the regulation of Ca^{2+} concentration, and for the release of protons (via $HCO_3^- \rightarrow CO_3^{2-} + H^+$) to liberate aqueous CO_2 in support of photosynthesis. They also propose the "Phosphate Extraction Mechanism" as a potential reason for the advent of biocalcification, which addresses the potential role that calcification plays in increasing the efficiency by which photosynthetic prokaryotes and eukaryotes can acquire phosphate in nutrient depleted environments.

The collection of articles presented in this Special Issue reveals the diversity of ongoing research into the effects of CO_2-induced acidification on calcareous biomineralization, while introducing some new questions and provocative hypotheses. Continued investigation of these concepts should advance understanding of the mechanisms of biocalcification and improve predictions of how future CO_2-induced changes in marine and freshwater systems will impact calcifying organisms, as well as the ecosystems they comprise, in the decades and centuries ahead. Although many questions remain unanswered, this collection of work supports the expanding body of evidence that the biological response to CO_2-induced acidification is more variable and complex than originally thought, requiring a more nuanced approach to exploring this important and fascinating process.

Author Contributions: Conceptualization of the Special Issue and this editorial, writing, editing: H.W., J.B.R. and S.S.D. All authors have read and agreed to the published version of the manuscript.

Funding: J.B.R. acknowledges salary support from MIT Sea Grant (NA18OAR4170105) while working on this editorial.

Acknowledgments: We acknowledge the support by MDPI staff, in particular Cynthia Li, during the course of making this Special Issue happen.

Conflicts of Interest: The authors declare no conflict of interest.

References

1. Carpenter, K.E.; Abrar, M.; Aeby, G.; Aronson, R.B.; Banks, S.; Bruckner, A.; Chiriboga, A.; Cortes, J.; Delbeek, J.C.; DeVantier, L.; et al. One-third of reef-building corals face elevated extinction risk from climate change and local impacts. *Science* **2008**, *321*, 560–563. [CrossRef]
2. Uthicke, S.; Momigliano, P.; Fabricius, K. High risk of extinction of benthic foraminifera in this century due to ocean acidification. *Sci. Rep.* **2013**, *3*, 1769. [CrossRef]
3. Fabricius, K.E.; Langdon, C.; Uthicke, S.; Humphrey, C.; Noonan, S.; De'Ath, G.; Okazaki, R.; Muehllehner, N.; Glas, M.S.; Lough, J.M. Losers and winners in coral reefs acclimatized to elevated carbon dioxide concentrations. *Nat. Clim. Chang.* **2011**, *1*, 165–169. [CrossRef]
4. Johnson, V.R.; Russell, B.D.; Fabricius, K.E.; Brownlee, C.; Hall-Spencer, J.M. Temperate and tropical brown macroalgae thrive, despite decalcification, along natural CO_2 gradients. *Global Chang. Biol.* **2012**, *18*, 2792–2803. [CrossRef]
5. Ries, J.B.; Cohen, A.L.; McCorkle, D.C. Marine calcifiers exhibit mixed responses to CO_2-induced ocean acidification. *Geology* **2009**, *37*, 1131–1134. [CrossRef]
6. Ries, J.B. A physicochemical framework for interpreting the biological calcification response to CO_2-induced ocean acidification. *Geochim. Cosmochim. Acta* **2011**, *75*, 4053–4064. [CrossRef]
7. Vogel, N.; Fabricius, K.E.; Strahl, J.; Noonan, S.H.C.; Wild, C.; Uthicke, S. Calcareous green alga *Halimeda* tolerates ocean acidification conditions at tropical carbon dioxide seeps. *Limnol. Oceanogr.* **2015**, *60*, 263–275. [CrossRef]
8. Guillermic, M.; Cameron, L.P.; DeCorte, I.; Misra, S.; Bijma, J.; de Beer, D.; Reymond, C.E.; Westphal, H.; Ries, J.B.; Eagle, R.A. Thermal stress reduces pocilloporid coral resilience to ocean acidification by impairing control over calcifying fluid chemistry. *Sci. Adv.* **2021**, *7*, eaba9958. [CrossRef] [PubMed]
9. Kroeker, K.J.; Kordas, R.L.; Crim, R.; Hendriks, I.E.; Ramajo, L.; Singh, G.S.; Duarte, C.M.; Gattuso, J.-P. Impacts of ocean acidification on marine organisms: Quantifying sensitivities and interaction with warming. *Glob. Chang. Biol.* **2013**, *19*, 1884–1896. [CrossRef]
10. Kroeker, K.J.; Micheli, F.; Gambi, M.C. Ocean acidification causes ecosystem shifts via altered competitive interactions. *Nat. Clim. Chang.* **2013**, *3*, 156–159. [CrossRef]
11. Kroeker, K.J.; Gambi, M.C.; Micheli, F. Community dynamics and ecosystem simplification in a high-CO_2 ocean. *Proc. Natl. Acad. Sci. USA* **2013**, *110*, 12721–12726. [CrossRef]
12. Stuhr, M.; Cameron, L.P.; Blank-Landeshammer, B.; Reymond, C.E.; Doo, S.S.; Westphal, H.; Sickmann, A.; Ries, R.B. Divergent proteomic responses offer insights into resistant physiological responses of a reef-foraminifera to climate change scenarios. *Oceans* **2021**, *2*, 281–314. [CrossRef]
13. Stuhr, M.; Meyer, A.; Reymond, C.E.; Narayan, G.R.; Rieder, V.; Rahnenführer, J.; Kucera, M.; Westphal, H.; Muhando, C.A.; Hallock, P. Variable thermal stress tolerance of the reef-associated symbiont-bearing foraminifera *Amphistegina* linked to differences in symbiont type. *Coral Reefs* **2018**, *37*, 811–824. [CrossRef]
14. Ninokawa, A.; Ries, J.B. Responses of freshwater calcifiers to carbon-dioxide-induced acidification. *J. Mar. Sci. Eng.* 2022, *submitted*.
15. Dodd, L.F.; Grabowski, J.H.; Piehler, M.F.; Westfield, I.; Ries, J.B. Ocean acidification impairs crab foraging behaviour. *Proc. R. Soc. Lond. B* **2015**, *282*, 20150333. [CrossRef]
16. Dodd, L.F.; Grabowski, J.H.; Piehler, M.F.; Westfield, I.; Ries, J.B. Juvenile Eastern Oysters more resilient to extreme ocean acidification than their mud crab predators. *Geochem. Geophys. Geosyst.* **2021**, *22*. [CrossRef]
17. Ries, J.B.; Cameron, L.P.; Reymond, C.E.; Bijma, J.; Büscher, J.V.; de Beer, D.; Guillermic, M.; Eagle, R.A.; Gunnell, J.; Müller-Lundin, F.; et al. Impacts of warming and acidification on coral calcification linked to photosymbiont loss and deregulation of calcifying fluid pH. *J. Mar. Sci. Eng.* 2022, *submitted*.
18. Eagle, R.A.; Guillermic, M.; De Corte, I.; Alvarez Caraveo, B.; Bove, C.B.; Misra, S.; Cameron, L.P.; Castillo, K.D.; Ries, J.B. Physicochemical control of Caribbean coral calcification linked to host and symbiont responses to varying pCO_2 and temperature. *J. Mar. Sci. Eng.* 2022, *in press*.

19. Khalil, M.; Doo, S.S.; Stuhr, M.; Westphal, H. Ocean warming amplifies effects of ocean acidification on skeletal mineralogy and microstructure in the asterinid starfish *Aquilonastra yairi*. *J. Mar. Sci. Eng.* 2022, *submitted*.
20. George, M.N.; O'Donnell, M.J.; Concodello, M.; Carrington, E. Mussels repair shell damage despite limitations imposed by ocean acidification. *J. Mar. Sci. Eng.* **2022**, *10*, 359. [CrossRef]
21. Scucchia, F.; Sauer, K.; Zaslansky, P.; Mass, T. Artificial intelligence reveals the 3D internal skeletal architecture in newly settled coral recruits: Insights into the effect of Ocean Acidification on coral biomineralization. *J. Mar. Sci. Eng.* **2022**, *10*, 391. [CrossRef]
22. Horvath, K.M.; Ries, J.B.; Castillo, K.D.; Westfield, I.T.; Armstrong, P.; Courtney, T. Next-century ocean acidification and warming both reduce calcification rate, but only acidification alters skeletal morphology of reef-building coral *Siderastrea siderea*. *Sci. Rep.* **2016**, *6*, 29613. [CrossRef]
23. Cohen, A.L.; McCorkle, D.C.; de Putron, S.; Gaetani, G.A.; Rose, K.A. Morphological and compositional changes in the skeletons of new coral recruits reared in acidified seawater: Insights into the biomineralization response to ocean acidification. *Geochem. Geophys. Geosyst.* **2009**, *10*, Q07005. [CrossRef]
24. Tambutté, E.; Venn, A.A.; Holcomb, M.; Segonds, N.; Techer, N.; Zoccola, D.; Allemand, D.; Tambutté, S. Morphological plasticity of the coral skeleton under CO_2-driven seawater acidification. *Nat. Comm.* **2015**, *6*, 7368. [CrossRef] [PubMed]
25. Hönisch, B.; Ridgwell, A.; Schmidt, D.; Thomas, E.; Gibbs, S.; Sluijs, A.; Zeebe, R.; Kump, L.; Martindale, R.; Greene, S.; et al. The geologic record of ocean acidification. *Science* **2012**, *335*, 1058–1063. [CrossRef] [PubMed]
26. Pomar, L.; Hallock, P.; Mateu Vicens, G. Why do carbonates exist? *J. Mar. Sci. Eng.* 2022, *in preparation*.

Journal of
Marine Science and Engineering

MDPI

Article

Mussels Repair Shell Damage despite Limitations Imposed by Ocean Acidification

Matthew N. George [1,2,*], Michael J. O'Donnell [1,2,†], Michael Concodello [3] and Emily Carrington [1,2]

1 Friday Harbor Laboratories, University of Washington, Friday Harbor, WA 98250, USA; mooseo@berkeley.edu (M.J.O.); ecarring@uw.edu (E.C.)
2 Department of Biology, University of Washington, Seattle, WA 98195, USA
3 Department of Biology, University of Rhode Island, Kingston, RI 02881, USA; mconcodello@gmail.com
* Correspondence: mngeorge@uw.edu
† Current Address: Department of Bioengineering, University of California, Berkeley, CA 94720, USA.

Abstract: Bivalves frequently withstand shell damage that must be quickly repaired to ensure survival. While the processes that underlie larval shell development have been extensively studied within the context of ocean acidification (OA), it remains unclear whether shell repair is impacted by elevated pCO_2. To better understand the stereotypical shell repair process, we monitored mussels (*Mytilus edulis*) with sublethal shell damage that breached the mantle cavity within both field and laboratory conditions to characterize the deposition rate, composition, and integrity of repaired shell. Results were then compared with a laboratory experiment wherein mussels (*Mytilus trossulus*) repaired shell damage in one of seven pCO_2 treatments (400–2500 μatm). Shell repair proceeded through distinct stages; an organic membrane first covered the damaged area (days 1–15), followed by the deposition of calcite crystals (days 22–43) and aragonite tablets (days 51–69). OA did not impact the ability of mussels to close drill holes, nor the microstructure, composition, or integrity of end-point repaired shell after 10 weeks, as measured by μCT and SEM imaging, energy-dispersive X-ray (EDX) analysis, and mechanical testing. However, significant interactions between pCO_2, the length of exposure to treatment conditions, the strength and inorganic content of shell, and the physiological condition of mussels within OA treatments were observed. These results suggest that while OA does not prevent adult mussels from repairing or mineralizing shell, both OA and shell damage may elicit stress responses that impose energetic constraints on mussel physiology.

Keywords: biomineralization; calcification; *Mytilus edulis*; *Mytilus trossulus*; predator-prey interactions

Citation: George, M.N.; O'Donnell, M.J.; Concodello, M.; Carrington, E. Mussels Repair Shell Damage despite Limitations Imposed by Ocean Acidification. *J. Mar. Sci. Eng.* **2022**, *10*, 359. https://doi.org/10.3390/jmse10030359

Academic Editors: Hildegard Westphal, Justin Ries and Steve Doo

Received: 7 January 2022
Accepted: 26 February 2022
Published: 3 March 2022

1. Introduction

Mytilid mussels are bivalve mollusks that perform ecologically important roles within marine ecosystems [1,2] and support an intercontinental fishery that accounts for 6.9% (1.2 million tonnes) of the USD 250 billion global aquaculture industry [3]. Mussels survive harsh conditions along coastal shores using bivalve shells that protect them from waves, predators, and desiccation [4]. Nearshore ecosystems are subject to substantial environmental variability, including fluctuations in seawater pH [5]. As a consequence of abiotic and biotic nearshore processes, mussels routinely endure exposure to acidified conditions (with respect to open ocean conditions), resulting in pH reductions of up to 1 unit for minutes, hours, or even days at a time [6–8]. Ocean acidification (OA), or the incremental decline in oceanic pH globally that results from the uptake of anthropogenic atmospheric pCO_2 by the ocean [9,10], is expected to intensify these processes, resulting in significant consequences for marine shelled organisms [11]. Given their ecological and economic importance, the ability of bivalves to build, maintain, and repair damaged shell under different pCO_2 conditions will ultimately determine the impact that OA has on entire ecological communities and aquaculture production [12,13].

The distribution of mussels within the intertidal zone is limited by wave forces [14] and predation [15,16]. Sea stars, crustaceans, birds, and snails (i.e., muricid gastropods) prey upon mussels by prying, peeling, cracking, and/or drilling shell to access the soft tissue within [17–21]. However, predation attempts on shelled mollusks are frequently unsuccessful, resulting in sublethal shell damage that must be quickly repaired to ensure survival [22,23]. Additionally, endolithic attack by bacteria or fungi [24] can also result in the formation of holes that expose the mantle cavity. Even a temporary breach in shell integrity can be a serious liability, as metabolites that escape the shell cavity quickly attract potential predators [25–27]. Shell breaches can also increase desiccation risk during low tide, stimulate an immune response [28], and hinder acid–base regulation [29]. To mitigate these risks, mussels and other mollusk species are capable of quickly repairing a shell injury by patching holes [30–32]. Using this strategy, mollusks seal off the mantle cavity within hours or days [33,34], although weeks or months may be required to regain the structural integrity of unrepaired shell [35–37].

Mussel shell is a complex biomaterial that is composed of three distinct layers. The outer shell is composed of a proteinaceous covering (periostracum) [38], the middle layer is composed of slender calcite fibers [39], and the innermost layer is composed of tablet aragonite crystals (nacre, 'mother-of-pearl') [40]. The outer mantle (marginal and pallial zones) produces both calcite and aragonite during shell growth, while the inner mantle (central zone) produces aragonite and is involved in shell thickening and repair [41]. In response to shell damage, the mantle extrudes a mixture of polysaccharides and glycoproteins into the extrapallial fluid, forming an organic membrane that covers the interior of the shell breach through a process that has been well described within mollusks [42,43]. A suite of conserved [44] and species-specific [45] shell matrix proteins (SMPs) then determine which mineral polymorph forms on the membrane by facilitating the nucleation, growth, and termination of calcium carbonate crystals (for a review, see [46]). Organic sheet synthesis has been observed to occur 20 days after shell injury in *Mytilus*, followed by the deposition of calcite crystals (29 days) and aragonite tablets (36 days) [47].

The effect of OA on shell formation in mussels has been well studied at the growth margin [11,48–50], particularly in the context of larval development [51,52] and in the presence of warming [53–55]. Aragonite saturation-state has been shown to be the most influential variable in juvenile mussel and oyster calcification [56]. While not extensively studied in adults, OA can impose metabolic constraints on the biomineralization process [57], resulting in changes in shell shape and thickness [58]. An environmentally induced metabolic constraint could be particularly relevant for shell repair, given the creation and maintenance of the organic matrix is costly [59,60]. OA can also impact shell formation through other direct and indirect mechanisms, including a decrease in carbonate ion ($CO^{2-}{}_{3}$) availability with declining pH [56,61] and pH-induced metabolic stress that disrupts the intercellular transport mechanisms that support the production of calcium carbonate [62]. However, uncertainty still exists as to the extent to which the processes that underlie shell repair mimic biomineralization at the shell margin [47] and whether the repair process in adults is subject to the same kinetic constraints as in early shell development [63].

Here we present data collected from a field experiment conducted in the rocky intertidal zone of Rhode Island, USA, on the blue mussel (*Mytilus edulis*; Linnaeus, 1758) and an OA laboratory experiment conducted in Washington, USA, on the pacific blue mussel (*Mytilus trossulus*; Gould, 1850). Our field observations and experiments serve to determine the frequency with which mussels sustain shell damage and provide a timeline of the shell repair process. These results were then used to inform laboratory mesocosm experiments wherein mussels repaired sublethal shell damage that breached the mantle cavity away from the shell margin under seven OA treatments (pCO_2 targets: 400–2500 µatm) for up to 10 weeks. The impact of environmental pCO_2 on the progression of the shell repair process, as well as the composition, microstructure, and strength of repaired shell, was then assessed through mechanical testing, SEM imaging, and µCT analysis.

2. Materials and Methods

Datasets presented here are the compilation of field and laboratory experiments undertaken on the Atlantic and Pacific coasts of the United States, utilizing two species of mytilid mussels, *M. edulis* and *M. trossulus*, respectively. Initial field observations of the frequency of shell repair in *M. edulis* populations were collected over three years (1998–2000) during monthly mussel bed surveys within the intertidal zone near Bass Rock and Black Point in Rhode Island, USA (41°24′17.4″ N, 71°27′30.5″ W; 41°23′42.4″ N, 71°27′47.9″ W), following previously described sampling methods [64,65]. The observation that mussels frequently sustained and repaired shell damage from boring predators (Figure 1) motivated field experiments during the summer of 2003. In field experiments, sublethal shell damage that breached the mantle cavity was induced in *M. edulis* living within bed populations at Black Point; mussels were then resampled over the course of two months to investigate the repair process (Figures 2 and 3). Results of these field experiments were subsequently used to inform a laboratory experiment investigating the effect of ocean acidification (OA; elevated seawater pCO_2, decreased pH) on the shell repair process in *M. trossulus*, conducted during the summer of 2012 at Friday Harbor Laboratories, located on San Juan Island, Washington, USA (48°32′46.9″ N, 123°00′36.5″ W).

M. trossulus and *M. edulis* are closely related sister taxa that naturally occur in sympatric populations along the eastern and western coasts of the United States. Due to their genetic similarity and systematic inclusion, along with the Mediterranean mussel (*M. galloprovincialis*), within the 'Mytilus edulis complex' [66,67], hybridization between the two species is common [68,69]. While members of the complex have similar growth rates and physiology [70], shell characteristics can vary. For example, *M. edulis* typically produces stronger, thicker shells than *M. trossulus*, with the magnitude of this species difference varying by site [71]. However, the material properties and composition of shell (e.g., Young's modulus, Vickers hardness, calcite/aragonite crystallographic orientation) within *M. edulis* and *M. trossulus* are not significantly different [72], suggesting their response to shell damage within this study is comparable.

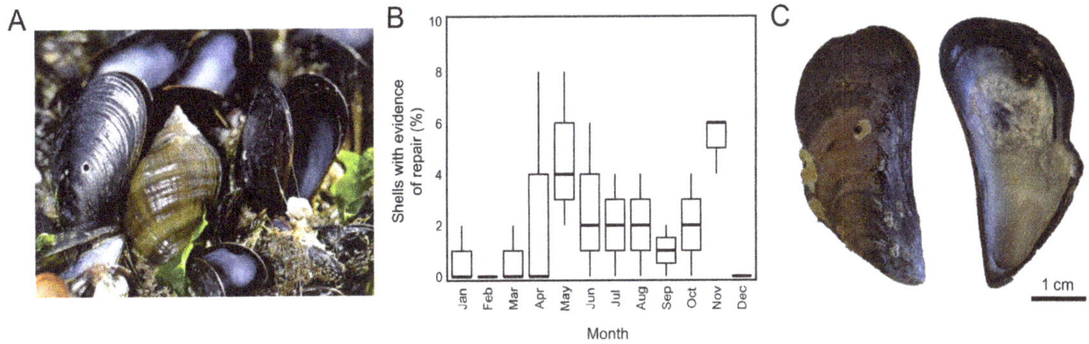

Figure 1. (**A**) Predation on mussels (*Mytilus edulis*) by predatory gastropods (*Nucella lapillus*; image credit: Luke Miller). (**B**) Mussels (*M. edulis*, n = 50 per sample) with evidence of repair of gastropod drill holes over 3 years of monthly field sampling in Rhode Island. (**C**) Exterior and interior view of a shell (*M. edulis*) with a repaired drill hole collected during field sampling within Rhode Island.

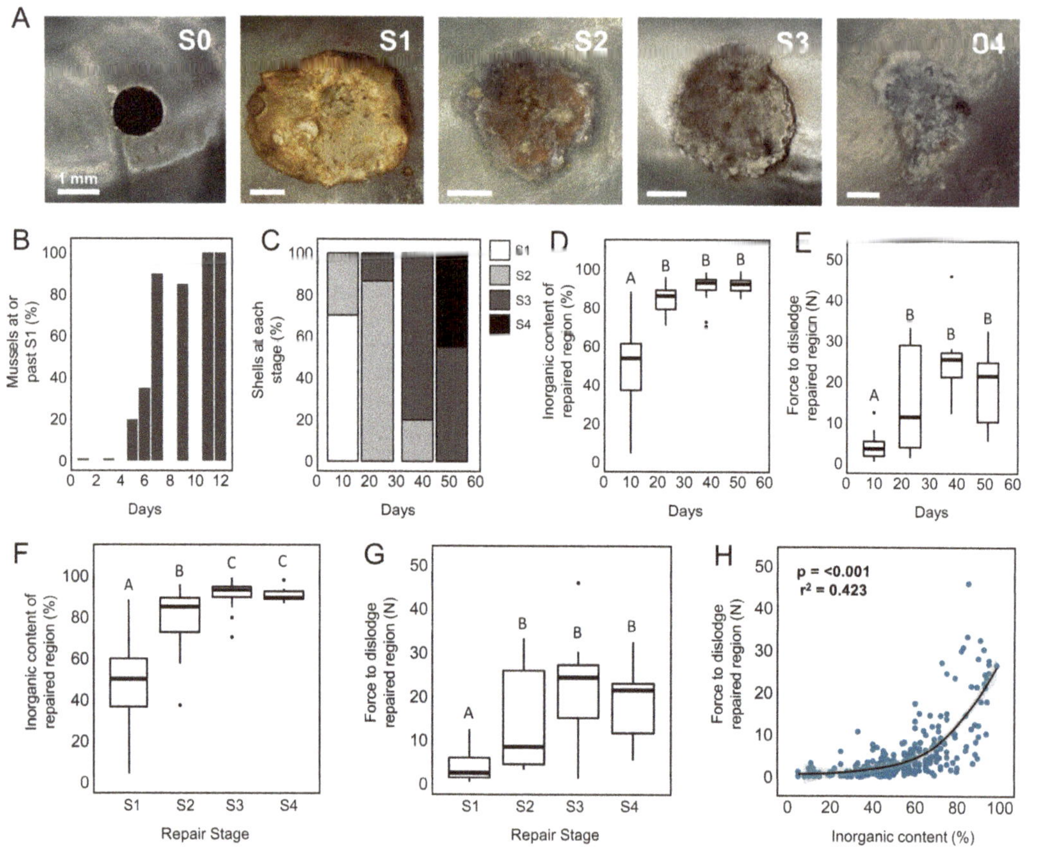

Figure 2. Time series of shell repair process. (**A**) Photographs of the interior of damaged mussel shells showing the four stereotypical repair stages (S1–S4), as defined in this study. (**B**) Proportion of mussels (*M. trossulus*, n = 25) within laboratory experiments that closed drill holes (reached or exceeded S1) over 12 days. Proportion of mussels (*M. edulis*, n = 15 per treatment) at each repair stage (**C**), the inorganic content of excised repaired shell (**D**), and the force required to dislodge repaired shell (**E**), from out-planted populations sampled over seven weeks in the intertidal zone. Summary of the inorganic content (**F**) and force to dislodge (**G**) repaired shell within each repair stage (pooled data from *M. edulis* and *M. trossulus*, n = 282). The relationship between the force required to dislodge repair patches and their inorganic content (**H**) across field and laboratory experiments using pooled data from both mussel species. Letters within panels indicate the results of tukey HSD post-hoc comparisons; groups that share a letter are not significantly different.

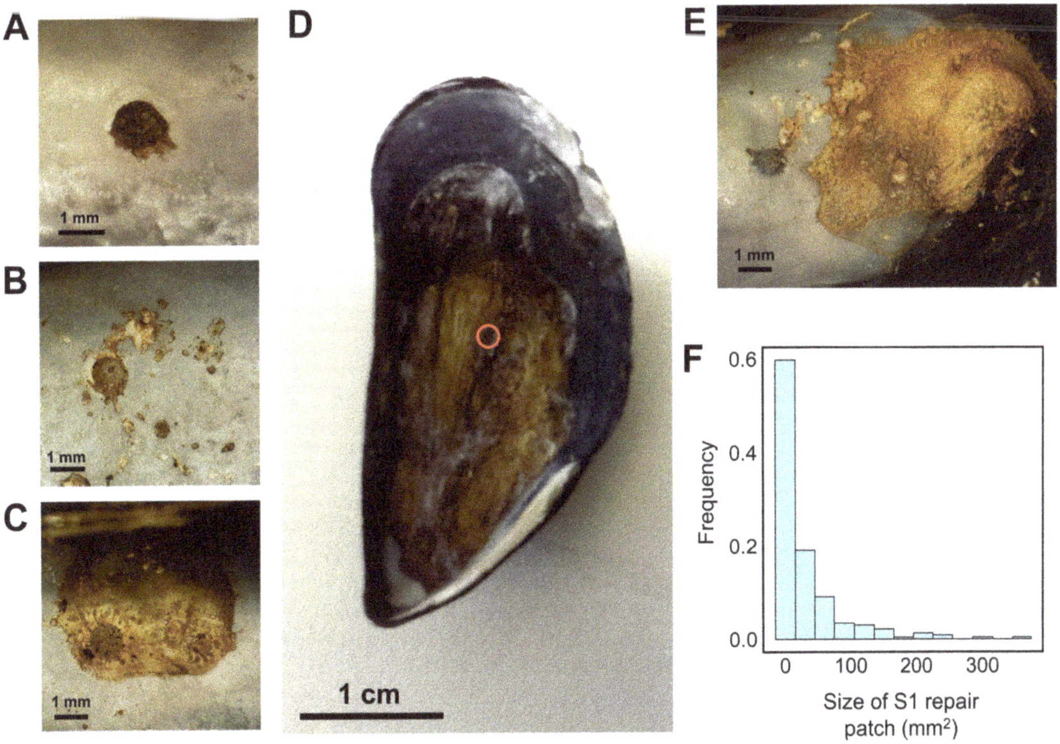

Figure 3. Examples of variable response to shell damage during Stage 1 (S1; *M. edulis*). Some mussels deposited organic matrix neatly within the shell defect (**A**), while others applied repair patches over a greater area (**B**,**C**). In rare cases, the organic matrix encompassed the entirety of the valve interior ((**D**), red circle indicates location of drill hole), while others produced matrix away from the shell injury altogether (**E**). The frequency distribution of repair patch size during S1 (**F**).

Throughout field and laboratory experiments, several measurement techniques remained consistent. Whenever a mussel was sampled, the shell length of the major valve axis was measured with vernier calipers (Wiha-41102, ±0.1 mm). Mussel condition and reproductive status were also commonly assessed by dissecting and separating gonad and somatic tissue and drying each at 60 °C until a stable weight was achieved. Condition index (CI) was defined as the ratio of total dry tissue mass to shell length cubed [73], while gonad index (GI) was defined as the ratio of the dry gonad mass to dry somatic tissue mass (Metler Toledo ML54, ±0.01 g) as previously described [64].

To assess the progression of shell repair through distinct repair stages, shells were dried at room temperature, photographed with a length standard, and qualitatively scored from 0 to 4. Shells that had no visible evidence of repair were assigned to Stage 0 (S0). Shells where organic matrix covered drill holes (or was present elsewhere) were considered to be at Stage 1 (S1). Shells at Stage 2 (S2) displayed a mixture of organic and mineral material, while Stage 3 (S3) was assigned when repairs were completely mineralized and rough in texture. Repaired shells at Stage 4 (S4) were considered to be visually similar to the surrounding shell material and resembled the surface shine, color, and texture of nacre. Examples of repaired shells at each stage are provided in Figure 2A.

The size of shell repair patches at each stage was quantified by photographing the shell interior and tracing the outline of the repaired region in Image J [74] to determine its cross-sectional area (±1 mm^2). The strength of repaired shell was assessed using an Instron

5565 material testing frame [75] fitted with a microindentation steel tip (diameter = 0.5 mm). Repairs were approached from the exterior of the shell at 10 mm min^{-1}, and the resulting maximum force required to dislodge the repaired region from drill holes was recorded (±0.01 N). This assay was not designed to be a comprehensive analysis of the material properties of repaired shell, but rather to approximate the effort a predator would have to exert to re-enter the mantle cavity. Finally, the inorganic content (%) of repaired shell was determined by removing newly deposited material from shell injuries and comparing their mass before and after incineration at 500 °C for 4 h [76].

2.1. Rhode Island: Field Experiment

The ability of *M. edulis* (shell length = 30–50 mm) to repair shell damage was first assessed within mussel bed communities located within the intertidal zone near Black Point, Rhode Island, USA. A 1 mm diameter hole was carefully drilled in the center of the right valve of mussels in situ without removing mussels from aggregations, using a drill stop of 1 mm to prevent injury to the internal tissues of the animal. Mussels chosen for inclusion in the experiment were at similar tidal heights (ranging from 0.5 m above and below MLLW) and within three meters of each other to ensure consistent wave exposure and environmental conditions. After sustaining sublethal shell damage, mussels were sampled over the course of two months (June–July 2003) approximately biweekly (10, 22, 38, and 51 days). Upon collection, animals were sacrificed and the CI and GI of each were determined; these metrics were compared with an initial sample of nearby mussels taken on the day shell damage was initiated. Shells were visually assessed for shell repair (qualitative score S0–S4), and the cross-sectional area of the repaired region (mm^2), the force required to dislodge the shell repair (N), and the inorganic content (%) of newly produced shell were measured using methods as previously described.

2.2. Washington State: Laboratory Ocean Acidification Experiment

M. trossulus (shell length = 30–50 mm) were collected from Argyle Creek, San Juan Island, WA, USA (48-31′12″ N, 123-00′53″ W), in March 2012. Upon collection, a subset of mussels was immediately sampled for initial field values of GI and CI. Shell damage was induced in the right valve of remaining mussels as previously described, and individuals were haphazardly placed in one of seven experimental mesocosms that ranged in target pCO_2 levels (400, 700, 1000, 1600, 1900, 2200, 2500 µatm) at 16 °C in the Ocean Acidification Environmental Laboratory (OAEL) located at Friday Harbor Laboratories, San Juan Island, WA, USA. Mussels were held in 1.5 L chambers with flow-through, UV-sterilized, and 0.2 µm filtered seawater. Chambers were cleaned three times weekly. Mussels were fed Shellfish Diet 1800 algal paste (Reed Mariculture, Campbell, CA, USA) at a daily rate of 5% of the estimated biomass within each chamber. Mussels were removed from each treatment over the course of a 2.5-month exposure at irregular intervals (8, 15, 22, 28, 43, 56, 69 days), and the GI, CI, and cross-sectional area of the repaired shell region, as well as the strength and inorganic content of repaired shell, were determined as previously described.

OA treatments were accomplished through dynamic injection of CO_2 using a pH-stat system, following the methods outlined by O'Donnell et al. (2013) [77]. Briefly, a Honeywell UDA2182 process controller and Honeywell Durafet III electrode [78] monitored the pH (uncertainty = ±0.10%) and temperature (uncertainty = ±0.63%) of each experimental mesocosm and added CO_2 to maintain the pH at a predefined setpoint calculated from target pCO_2 levels using CO2calc [79]. pH electrodes were calibrated to the total scale using spectrophotometric pH (Ocean Optics USB4000; Ocean Insight, Toms River, NJ, USA) and were compared to treatment conditions every 3–4 days to ensure the correct calibration was maintained. The salinity of each treatment was measured daily using a sensION 5 conductivity meter (Hach Company, Loveland, CO, USA; uncertainty = ±0.33%). Total alkalinity (AT) was measured using SOP 3b from Dickson et al. (2007) every 3–4 days (uncertainty = ±0.33%).

The relationship between AT and salinity established over the course of two years at our field station (AT = 38.856 * Salinity + 916.43, R2 = 0.95) was used to estimate AT in each mesocosm; results obtained by this method were found to be within ±0.4% of measured AT values. From estimates of AT and measurements of pH, temperature, and salinity, we calculated the pCO_2 (μatm), CO_3 (μmol kg^{-1} SW), HCO_3 (μmol kg^{-1} SW), aragonite saturation state (Ω_{ar}), and calcite saturation state (Ω_{ca}) of each treatment. The uncertainty associated with each calculated parameter was determined using a Monte Carlo analysis (i = 10,000), sampling the random, normal distribution of measurement uncertainty associated with each pH, AT, temperature, and salinity measurement and propagating them through each calculation. The resulting propagated uncertainty was combined with treatment variability (1 S.D.) by taking the square root of the sum of squares (reported as total uncertainty (U_{total})), following published recommendations [80].

2.3. SEM and μCT Imaging

Scanning electron micrograph (SEM) images of the surface of repaired shells (*M. trossulus*) at each repair stage (S1–S4; 400 μatm), as well as S4 repaired shell produced within the most extreme OA treatment (2500 μatm) over 10 weeks, were taken using a Sirion XL30 Field Emission SEM (FEI/Philips, Hillsboro, OR, USA). Shells were rinsed with dH_2O, air dried, flushed with nitrogen, and sputter coated with carbon (10 nm layer thickness) prior to being imaged at 25×, 1000×, and 3000× magnification at 5 kV. The elemental composition of the surface of samples was further determined using energy-dispersive X-ray (EDX) spectroscopy using a 100 μm^2 scan area (n = 3 mussels per group).

Microtomography (μCT) scans of shells from mussels (*M. trossulus*) included in OA treatments were taken using a Skyscan 1076 scanner (Bruker, Billerica, MA, USA), imaging shells in 35 μm slices at 45 kV. Three-dimensional image reconstruction was performed in NRecon (Micro Photonics Inc., Allentown, PA, USA), with further rendering in Drishti [81]. A 1 mm diameter cylinder was placed on the drill hole of each shell and used to record the mean and maximum grayscale values of scan slices in aggregate. Grayscale values were compared with those of unrepaired shell 1 mm away from the drill hole.

2.4. Statistical Analyses

All statistical analyses were performed in R (Version 4.1.0) using the RStudio IDE (Version 1.4.1717). When applicable, analysis of covariance (ANCOVA) was used to investigate differences in response variables to the duration of exposure (days) and magnitude (pCO_2 targets) of OA treatments. During model construction, the assumptions of normality and homoscedasticity were assessed using the Shapiro test and a visual assessment of Q-Q and residual-fitted plots. To achieve normality, the Johnson transformation was used when necessary [82]. When response variables were expressed as proportions, the logit transformation (log of odds ratio) was used. For significant effects (α = 0.05), the agricolae package was used to perform pairwise comparisons of groups using the Tukey HSD post hoc test [83]. For the comparison of qualitative repair scores, the distribution of mussels within each stage was compared with a chi-squared test, using the 400 μatm treatment values as the expected values.

3. Results

3.1. Rhode Island: Field Experiment

Evidence of gastropod predation within mussel beds varied significantly during monthly field sampling of intertidal sites, with as many as 8% of mussels (*M. edulis*, n = 50, 1998–2001) within bed populations carrying shell damage in a given month (Figure 1B). When shell damage was intentionally induced in a subset of individuals within a population, mussels progressed steadily through each repair stage (S0–S4) over a 51-day period (Figure 2A,C). Ten days after shell damage was induced, 70% of mussels had entered the first stage of shell repair (S1) and successfully closed drill holes by applying an organic membrane over the opening (examples from *M. trossulus* provided in Figure 3). These results matched laboratory assays wherein it took *M. trossulus* ($n = 25$) 11 days for all mussels to reach S1 (Figure 2B). Following the closure of the shell opening, 86.6% of mussels were at S2 after 22 days, and 80% were at S3 after 38 days (Figure 2C).

Significant changes in the material and biomechanical properties of repaired shell were observed as mussels progressed through each repair stage. The inorganic content ($p < 0.001$, Figure 2D) and force required to dislodge repaired shell material ($p < 0.001$, Figure 2E) significantly increased as mussels (*M. edulis*) remained within the intertidal zone after shell injury by +83% and +346% (comparing 10 to 51 days), respectively. For both measured parameters, the hardening of the repaired region corresponded with the transition from S1 to S2 (Figure 2C–E). The relationship between repair stage and the physical properties of repaired shell was further validated by pooling data from field experiments (*M. edulis*) and laboratory studies discussed in the following section (*M. trossulus*); repair stage was positively correlated with inorganic content ($p = 0.012$; Figure 2F) and force ($p = 0.032$; Figure 2G). The strength of repaired shell and inorganic content were also positively correlated (loess regression) with each other when compared across both species, with inorganic content explaining 42% of the variance observed in force ($p < 0.001$, Figure 2H).

At the end of the field experiment, the appearance of repaired shell resembled that of surrounding shell, with all mussels proceeding to at least S3 after 51 days; in this end-point population, 45% of repairs were in S4 (Figure 2C). However, mechanical testing demonstrated that repaired shell at stages S2, S3, and S4 required similar forces to dislodge repair patches (Figure 2E), indicating that perhaps more time is needed to produce a material with a similar structural integrity to undamaged shell.

3.2. Washington State: Ocean Acidification Shell Repair Experiment

Laboratory experiments employed seven OA treatments, with measured pH values ranging from 7.29 to 7.95 (total scale, Table 1) and calculated pCO_2 levels ranging from 483 to 2458 μatm (Table 2). OA did not significantly affect whether mussels were able to repair damaged shell, with no observed impact of pCO_2 on the proportion of mussels that fully mineralized repaired shell (reached S3 or S4) after 4 weeks ($p = 0.53$, Table S1). All mussels closed drill holes irrespective of treatment, with no impact of pCO_2 ($p = 0.64$) or length of treatment exposure (time, $p = 0.57$) on the size of the S1 repair patch (Table S3). Repair patches were generally proportional to the degree of shell damage, neatly covering the drill hole in 60% of cases (Figure 3F). However, significant overgrowth of the repair patch did occur, resulting in organic matrix deposition within the entire valve interior (Figure 3D) and repair away from the shell defect in rare instances (Figure 3E).

Table 1. Measured seawater carbonate parameters during OA treatments and their respective variability (±1 SD). Measurement uncertainties for each parameter were as follows: temperature (T; 0.63%), salinity (S; 0.33%), pH (0.13%), and total alkalinity (A_T; 0.19%).

pCO_2 Target	T (°C)	Salinity	pH (total)	A_T (μmol × kgSW)
400	15.8 ± 0.1	30.0 ± 0.2	7.95 ± 0.03	2079 ± 7
700	16.1 ± 0.5	29.9 ± 0.3	7.77 ± 0.02	2083 ± 8
1000	15.9 ± 0.2	30.2 ± 0.1	7.64 ± 0.02	2080 ± 10
1600	16.0 ± 0.3	30.1 ± 0.2	7.40 ± 0.02	2086 ± 7
1900	16.0 ± 0.2	30.0 ± 0.1	7.38 ± 0.06	2080 ± 6
2200	16.0 ± 0.4	29.8 ± 0.2	7.31 ± 0.03	2078 ± 5
2500	15.9 ± 0.1	30.4 ± 0.3	7.29 ± 0.03	2090 ± 9

Table 2. Calculated seawater parameters over the course of OA treatments and their respective uncertainties. The total uncertainty (U_{total}) for each calculated parameter is reported as the combination of propagated measured uncertainties as reported in Table 1 and the variability of each parameter over the course of each experiment.

pCO_2 Target	pCO_2 (µatm)	CO_3 (µmol × kgSW)	HCO_3 (µmol × kgSW)	Ω_{Ar}	Ω_{Ca}
400	483 ± 64	110 ± 17	1807 ± 38	1.74 ± 0.26	2.73 ± 0.40
700	769 ± 100	77 ± 13	1892 ± 31	1.21 ± 0.22	1.90 ± 0.32
1000	1062 ± 140	58 ± 10	1939 ± 27	0.91 ± 0.16	1.43 ± 0.25
1600	1652 ± 215	39 ± 7	1986 ± 21	0.62 ± 0.11	0.97 ± 0.18
1900	2009 ± 372	34 ± 7	2000 ± 22	0.53 ± 0.11	0.82 ± 0.18
2200	2365 ± 317	28 ± 5	2013 ± 19	0.44 ± 0.08	0.69 ± 0.13
2500	2458 ± 340	27 ± 7	2016 ± 20	0.43 ± 0.09	0.67 ± 0.14

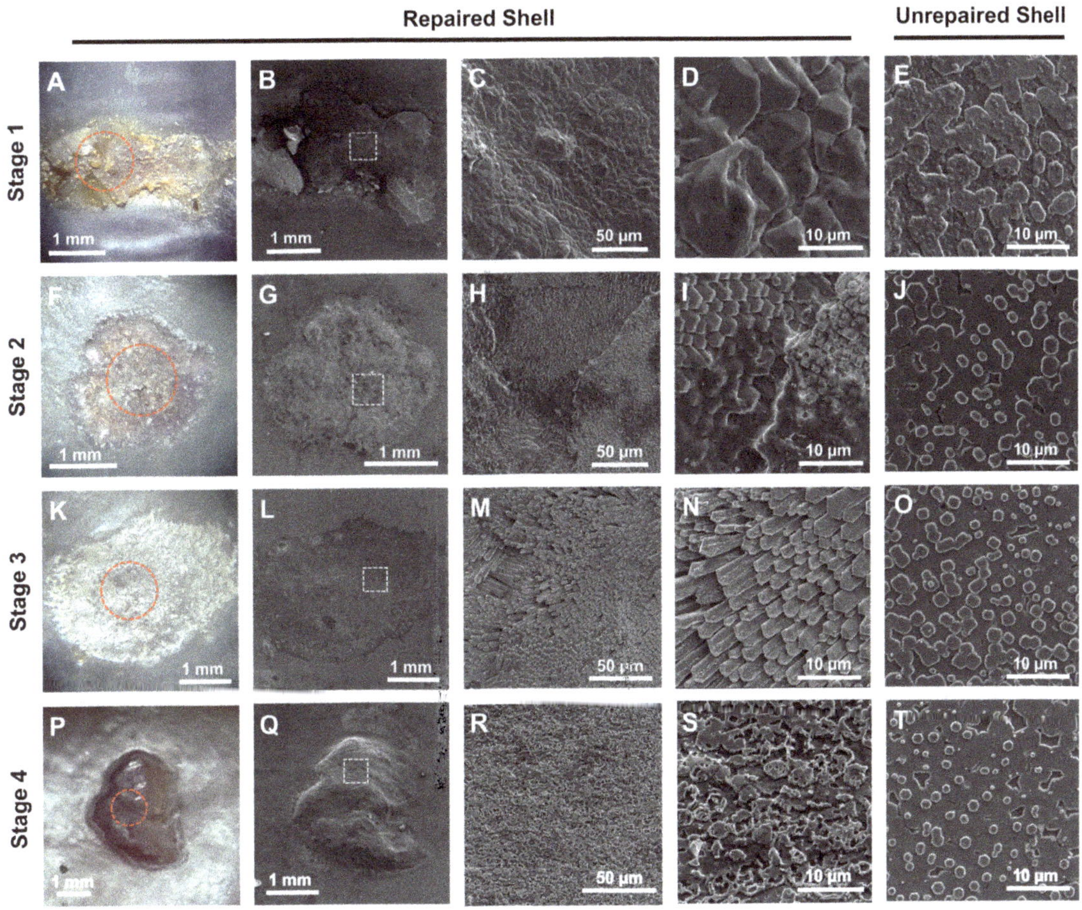

Figure 4. Light and SEM images of repaired drill holes at Stage 1 (S1; (**A–E**)), Stage 2 (S2; (**F–J**)), Stage 3 (S3; (**K–O**)), and Stage 4 (S4; (**P–T**)), from the valve interior of mussels (*M. trossulus*) exposed to 400 µatm pCO_2 for 15, 28, 43, and 69 days, respectively. Red circles represent the location of drill holes within light microscopy images; white boxes are the scan area used for magnified SEM images. Images for each repair stage were each taken from the same mussel, image magnification for SEM images increases from left to right within each row (25×, 1000×, 3000×). The last panel within each row is an image of unrepaired shell away from the repair within the same individual (3000×).

SEM imaging paired with EDX analysis of the interior surface of repaired shell produced by mussels within the 400 μatm treatment demonstrated that significant differences in shell microstructure (Figure 4) exist across repair stages. μCT imaging of shells at S3 and S4 also suggested that repaired shell had similar grayscale values to unrepaired shell (but not throughout), appeared thinner in cross-section, and was often anchored to the interior shell away from the drill hole edge (Figure 5C,D). No evidence of membrane mineralization was observed at S1 (Figure 4A–D), while a mixture of protein and short calcite fibers was observed at S2 (Figure 4F–I). At S3, the entire repair was made up of elongated calcite fibers (Figure 4K–N), which were covered by a layer of irregularly shaped aragonite tablets in S4 (Figure 4P–S). Tablets formed in S4 remotely resembled the nacreous layer present in unrepaired shell (Figure 4E,J,O,T), although with a pock-marked surface and an irregular orientation. A significant difference between the surface elemental composition of repaired shell and that of unrepaired shell was only observed in stages S1 ($X^2{}_{8100} = 192$, $p < 0.001$) and S2 ($X^2{}_{8100} = 7413$, $p < 0.001$) (Figure 7C).

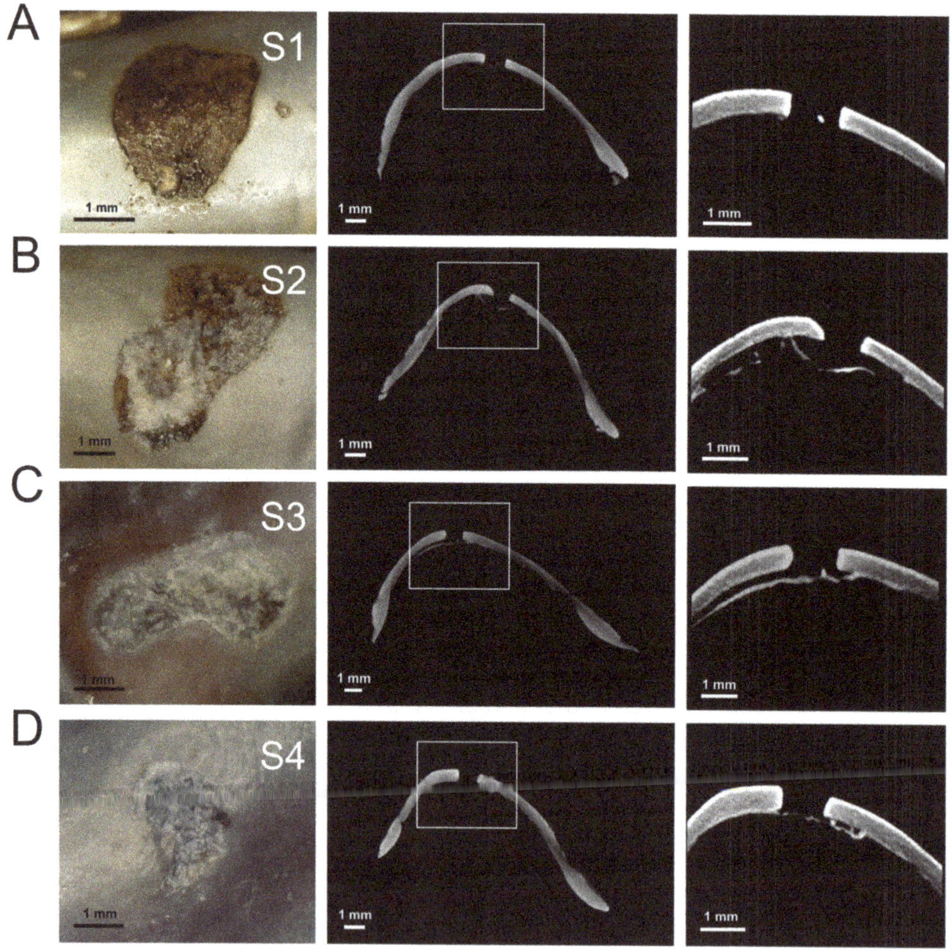

Figure 5. Representative photographs and μCT images of repaired drill holes at Stage 1 (S1; (**A**)), Stage 2 (S2; (**B**)), Stage 3 (S3, (**C**)), and Stage 4 (S4, (**D**)). Images are from mussels within laboratory experiments (*M. trossulus*; 400 μatm pCO_2).

While OA did not prevent mussels from closing shell injuries or mineralizing repaired shell, the severity of pCO_2 exposure and the time spent within treatments significantly impacted the inorganic content (OA: $p < 0.001$, time: $p < 0.001$; Table S3, Figure 6B) and the force required to dislodge repair patches (OA: $p < 0.001$, time: $p = 0.02$; Table S3, Figure 6C). Similar results were observed when analyses were constrained to only include mussels after 10 weeks within treatments (end-point only). After 10 weeks of OA exposure, significantly fewer animals reached S3 or S4 in treatments where pCO_2 was greater than 1500 µatm (Figure 6D; 400 µatm control). However, while pCO_2 did have a significant effect on the inorganic content of repaired shell ($p = 0.013$, Table S2, Figure 6E), no effect was observed on the force required to dislodge repair patches ($p = 0.263$, Table S2, Figure 6F). Similarly, no effect was observed on the surface elemental composition of S4 repaired shell produced under exposure to 2500 µatm pCO_2 for 10 weeks when compared with unrepaired shell from the same animal ($X^2{}_{8100} = 0.87$, $p = 0.99$) or conspecifics that produced S4 shell in the 400 µatm treatment ($X^2{}_{8100} = 1.99$, $p = 0.98$; see Figure 7C). Moreover, no effect of OA was observed on the mean ($p = 0.85$, Figure 8C) or maximum ($p = 0.56$, Figure 8D) grayscale values approximating the shell density of the repaired region collected from µCT scans (Table S4).

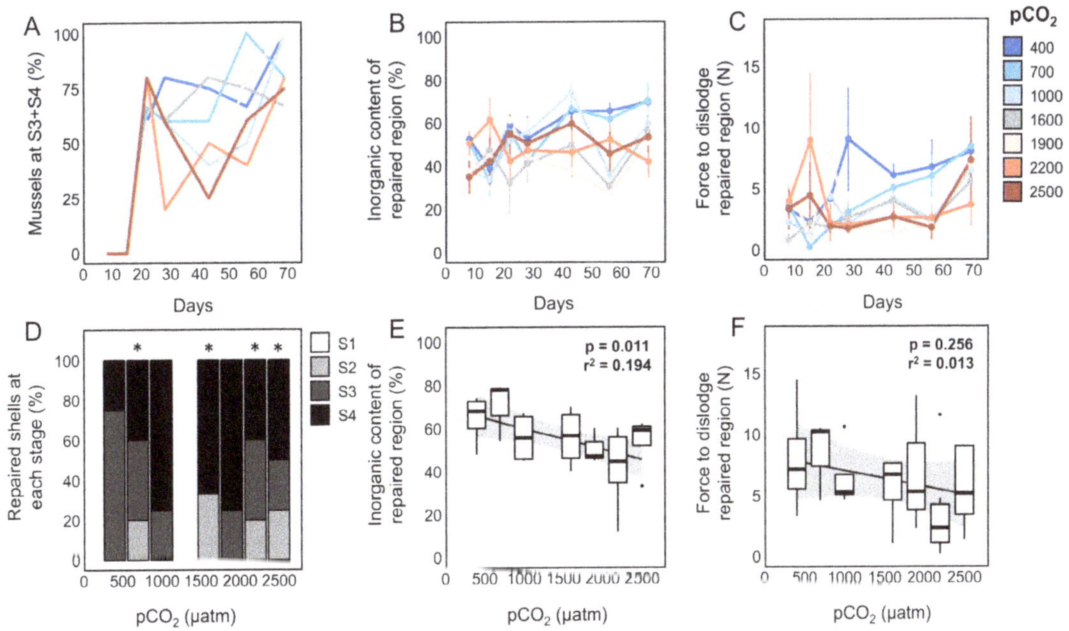

Figure 6. (**A**) Proportion of mussels (*M. trossulus*) that produced mineralized shell (reached S3 or S4) in response to shell damage within each OA treatment. (**B**) Inorganic content of excised repaired shell from mussels within each OA treatment. (**C**) The force required to dislodge repaired regions produced in each OA treatment. Proportion of mussels at each repair stage (**D**), the inorganic content of repaired regions (**E**), and force to dislodge repaired regions (**F**) after 10 weeks within each OA treatment. Data are from 4–8 mussels per treatment per time point. Asterisks mark treatments that were statistically different from the 400 µatm control.

The condition ($p < 0.001$) and gonad ($p = 0.008$) indices of mussels universally decreased over 10 weeks under laboratory conditions (Table S5, Figure S2A,D). Mussel condition ($p = 0.017$), but not reproductive condition ($p = 0.814$), was significantly affected by pCO_2, and no interaction with time in treatment was detected ($p = 0.645$, Table S3). When comparing the initial and final conditions and gonad indices under experimental conditions, a significant impact of pCO_2 on CI ($p < 0.001$) and GI ($p = 0.012$) was observed, with no decrease in either metric observed in field populations over the same time period (Table S5, Figure S1). However, neither CI nor GI was correlated with the force to dislodge repaired regions (CI: $p = 0.435$; GI: $p = 0.690$) or their inorganic content (CI: $p = 0.989$; GI: $p = 0.619$), with no observed clustering observed with pCO_2 treatment (Figure S2).

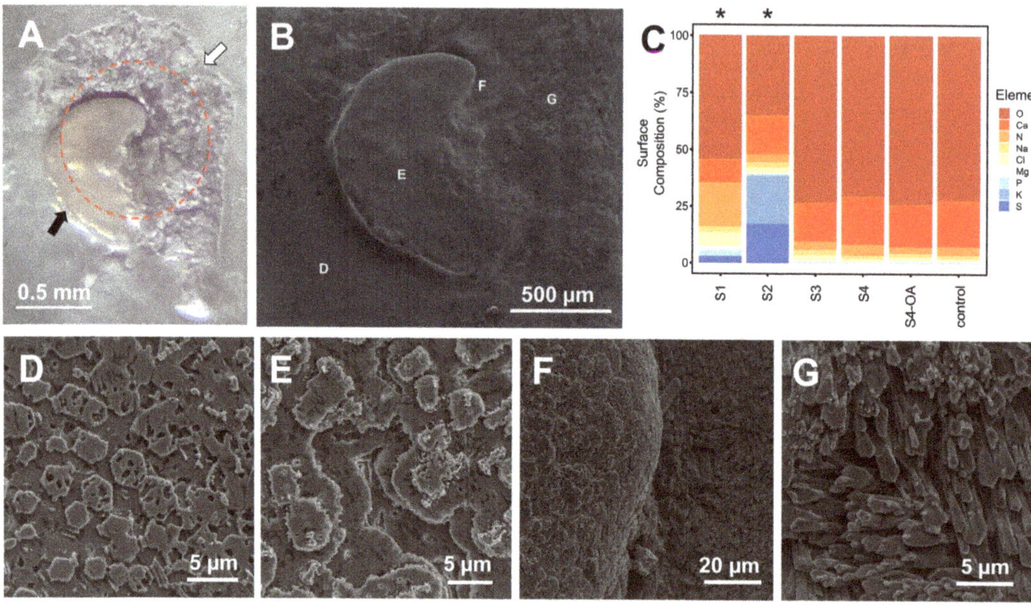

Figure 7. Light (**A**) and SEM (**B**,**D–G**) microscopy images of a repaired drill hole ((**A**), red circle) from the valve interior of a mussel (*M. trossulus*) exposed to 2500 μatm pCO_2 for 10 weeks, with evidence of repaired shell at both Stage 3 (S3; white arrow) and Stage 4 (S4; black arrow). (**C**) Elemental composition (%) of the interior valve shell surface at each repair stage (S1–S4; 400 μatm pCO_2), S4 shell produced under ocean acidification for 10 weeks (S4-OA; 2500 μatm pCO_2), and unrepaired shell (control) as determined through energy-dispersive X-ray (EDX) spectroscopy analysis (100 μm^2 scan area; $n = 3$ mussels per group). SEM images of unrepaired (**D**) and S4 repaired shell (**E**) confirmed the presence of a nacreous layer, although with different surface topographies. (**F**) SEM image of the transition between S4 (left) and S3 (right) repaired shell. (**G**) SEM image of S3 repaired shell resembling the calcite crystals commonly found in the oblique prismatic layer. Letters D–G within panel B indicate the locations of images within panels D–G. Asterisks indicate stages where the elemental composition was statistically different from that of the control.

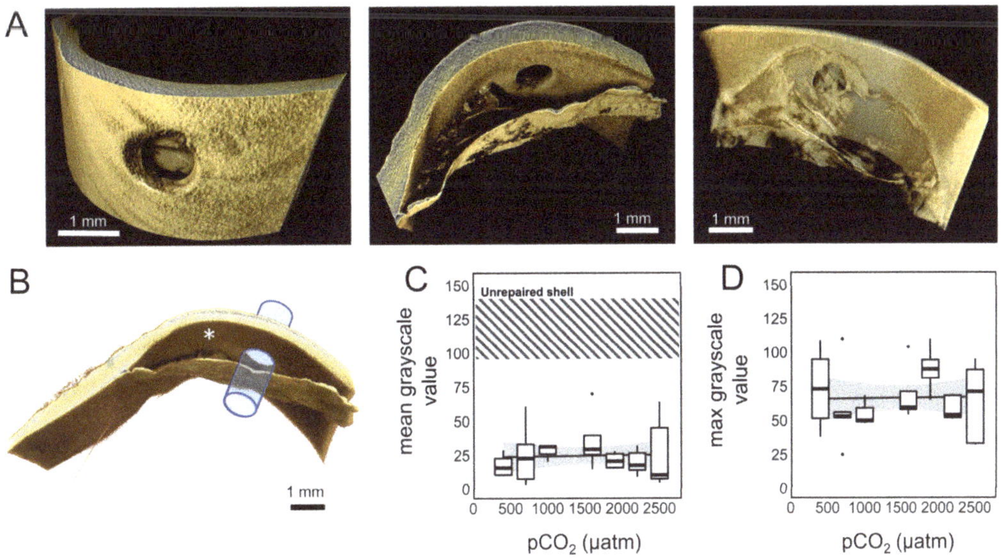

Figure 8. (**A**) The 3D rendering of a drill hole and deposited shell material constructed from µCT scan slices. Images represent three perspectives of the same shell repair from a single mussel (*M. trossulus*) held within the 400 µatm pCO_2 for 10 weeks. (**B**) Cylindrical volume used for analysis; asterisk marks approximate location used for control measurements of unrepaired shell. The mean (**C**) and maximum (**D**) grayscale values within sampled cylinders after 10 weeks within OA treatments (n = 4–8 mussels per treatment).

4. Discussion

Here we describe the process of shell repair in two mytilid mussel species following sublethal shell damage that penetrated the mantle cavity away from the shell margin, as well as the effect of ocean acidification (OA) on the deposition rate, structure, composition, and integrity of repaired shell. In both field and laboratory assays, mussels began to mineralize shell injuries within weeks, with repaired shell transitioning through repair stages defined by distinct changes in color, texture, and surface microstructure (S0–S4; Figures 2A and 4). Following damage, an organic membrane first covered the damaged area (days 1–15), followed by the deposition of calcite crystals (days 22–43) and aragonite tablets (days 51–69). Repair stage was also positively correlated with inorganic content, structural integrity, and calcium composition (Figures 2 and 7C). Considering end-point samples only, the ability of mussels to seal and mineralize shell breaches was not impacted by environmental pCO_2 (Table S1, Figure 6A), and no effect of OA was observed on the strength (Figure 6F), elemental composition (Figure 7C), microstructure (Figure 7D–G), or µCT grayscale values (Figure 8) of repaired shell after 10 weeks under laboratory conditions. However, when mussels from all time points across treatments were included in analyses, significant interactions between pCO_2, the length of exposure to treatment conditions, repair strength, repair inorganic content, and mussel physiological condition were observed (Table S3, Figure 6). These results suggest that, while OA (up to 2500 µatm) does not prevent mussels from repairing shell injuries, both pCO_2 exposure and shell damage likely elicit stress responses that impose energetic constraints on mussel physiology.

During S1, an organic membrane quickly covered drill holes. Incineration with muffle furnace (Figure 2D), µCT imaging (Figure 5), and EDX composition analysis (Figure 7C) confirmed that the membrane was not mineralized and did not have similar µCT grayscale values or a mineral composition that resembled any other stage. Membranes were also anchored to the shell around, rather than within, drill holes (Figures 5 and 8A). While a full characterization of the components of the shell-repair membrane in *Mytilus* remains to be undertaken, proteomic analyses of shell [84–86] and the extrapallial fluid [87,88] in multiple mytilid species suggest that the membrane is composed of a variety of polysaccharides, glycoproteins, and shell matrix proteins (SMPs). While only a minor component of the shell matrix, SMPs play an important role in $CaCO_3$ nucleation, growth, and polymorph determination [89,90], and their expression has been shown to be induced in the central mantle following shell injury [47]. Further, transcriptional analysis of the mantle following shell damage within *M. edulis* has identified transcripts that encode proteins with domains found in shell matrix proteins (SMPs), immune response proteins, and other proteins involved in biomineralization [45].

Significant variability was observed in the initial response to shell injury across mussels sampled, irrespective of species or pCO_2 treatment. Organic membranes typically covered drill holes but varied widely in their size and shape (Figure 3A–E); 10% of mussels produced a patch 100× greater than the drill hole diameter (Figure 3F), and in rare instances, organic matrix was deposited away from shell damage altogether (Figure 3E). To our knowledge, variability in the localization of the repair process of this magnitude has not been previously reported. One possible explanation for this variation could be that, while great lengths were taken to standardize the depth with which drill holes were generated, variation in shell thickness may have resulted in different degrees of tissue damage. Alternatively, nonlocalized repair could be the result of the nonspecific deposition of shell matrix proteins (SMPs) that act as nucleation sites during calcite and aragonite formation [45,47,91]. While it is likely that the transcriptional response of mantle tissue to shell damage will be greatest next to the shell injury [31], the factors that govern how SMPs are deposited onto damaged shell after extrusion into the extrapallial fluid remain unclear. The nonlocalized and oversized repair observed within some mussels within this study could point to this process being more generalized than previously thought, or mediated by some yet unknown property of the shell interior.

During shell biomineralization, calcium carbonate precipitation onto the organic matrix acts as a precursor to aragonite formation [30,92]. In this study, calcite filaments developed during S2 and became ubiquitous and elongated during S3 (days 22–43, Figure 4L–N), resembling the calcitic fibrous layer observed in *M. galloprovincialis* as described by Checa et al. (2014) [39]. This layer was then laminated with aragonite tablets (days 51–69; see Figure 4Q–S) that had a pock-marked, irregular surface topography when compared with unrepaired shell (e.g., see Figure 4T). Irrespective of individual variation in organic membrane size, the strength, inorganic content, and elemental calcium composition of repaired shell increased as it transitioned through each stage (Figures 2, 5 and 7C). As membranes were mineralized, the force required to dislodge repair patches from the shell exterior also increased, but did not approach that of unrepaired shell (Figure 2E; flat plate compression testing, George et al., unpublished data). The impact of shell repair on the mechanical strength of shells has been investigated in a number of mollusk species. In studies of gastropods that employed flat plate compression testing, mechanical stress produced cracks that propagated along growth lines rather than repaired shell, resulting in no impact on shell integrity [36,93]. Similarly, *M. californianus* has been found to repair shell fatigue damage, with repaired shell often exceeding the strength of unrepaired shell after 4 weeks [37]. However, these results are not comparable as the mechanical testing employed in this study was designed to estimate the effort required of a secondary predator to reach the mantle cavity through the repaired material. Future studies would benefit from employing multiple mechanical testing strategies to ascertain whether repaired drill holes impact the ability of shell to protect from different predators.

The repair chronology observed in this study is in agreement with other investigations of shell repair in a number of bivalve species [30,32,94]. Similarly, a study by Hüning et al. (2016) that employed a drill-based shell damage assay in *M. edulis* described three repair stages in which a repair membrane was evident at day 20, calcite crystal accumulation after 29 days, and aragonite tablet formation after 36 days [47]. While dividing the repair process into three stages may accurately describe the transition from a protein membrane to the two mineral polymorphs commonly seen during biomineralization, significant differences in the inorganic content (Figure 2F), microstructure (Figure 4F–I), and elemental composition (Figure 7C) were observed in this study during the transition between membrane formation and mineralization (S2). These results suggest that future studies would benefit from a more granular investigation of the repair process to accurately characterize the material and mechanical differences of shell within and between repair stages.

The morphology [58], chemical composition [95], microstructure [57,96], and mechanical properties [54,97] of mussel shells have been shown to be impacted by ocean acidification. OA has been proposed to impact shell formation through both direct and indirect mechanisms, including a decrease in carbonate ion ($CO_3{}^{2-}$) availability with declining pH [56,61], pH-induced metabolic stress that disrupts the intercellular transport mechanisms that support the production of calcium carbonate [62], decreased calcification rates [57], and an increase in the cost of calcification [98]. In this study, *M. trossulus* repaired damaged shell within seven pCO_2 treatments ranging from 400 to 2500 µatm for 10 weeks. No direct impact of OA was observed on the timing of the shell repair process, and all mussels reached S1 after 22 days regardless of OA treatment (Figure 6A). Looking at end-point samples (69 days of OA exposure), there was also no evidence of an impact of OA on the strength (Figure 6F) or µCT grayscale values (Figure 8) of repaired shell; a significant impact on the inorganic content was observed (Figure 6E). Mussels were able to reach S4 and produce aragonite tablets in the most extreme OA treatment (2500 µatm); tabloid structures were similar in both structure and composition to unrepaired shell produced within the acidified and control treatments (Figures 4 and 7). When all time points from all treatments were included, significant interactions between pCO_2 and the time spent in each OA treatment was observed for both the strength (force to dislodge) and inorganic content of repaired shell (Table S1), as well as a trend of more mussels remaining in S2 after 10 weeks in high pCO_2 treatments (Figure 6D).

Shell production is an important and cost-intensive process, with estimates that biomineralization can account for up to 22.3% of the energy budget of *M. trossulus* [99], with an estimated cost of 29 J mg^{-1} of organic matrix [59,60]. However, there is also substantial evidence that adult mussels can produce shell under physiologically stressful conditions, and many species persist in upwelling zones where CO_2-rich waters can lead to calcium carbonate saturation states well below 1 [100]. Subsequent observations of the total calcium carbonate production of mussel beds within these regions also suggest that the degree to which OA impacts shell production strongly depends on habitat food density (particulate organic carbon (POC)) [98,101] and pales in comparison to the effect of warming [54]. In the context of shell repair, robust expression of SMP candidate genes in the central mantle has also been shown to be maintained under exposure to 4000 µatm for 8 weeks [102]. In our laboratory experiment, all mussels were fed 5% of their wet body mass in algae daily, delivered at a concentration of 3000–10,000 cells mL^{-1} with peristaltic pumps at regular intervals. This amount of food is consistent with previous studies where mussels have maintained and even gained tissue mass over the course of months [77]. However, the condition of mussels within our experiment, as denoted by the ratio of grams of dried tissue to shell length cubed (CI), decreased as both a function of pCO_2 treatment ($p = 0.008$) and time ($p = 0.001$) with an interaction that was also significant ($p = 0.030$, Table S5). As a point of comparison, wild populations of mussels that did not undergo shell damage over this same time period did not experience a significant decrease in either physiology condition or gonad index (Table S5, Figure S1). To address this issue, future studies would benefit from the inclusion of additional controls to tease apart the energetic demand placed on mussels by the shell repair process, OA, and their interaction.

Our results suggest that mussels within the intertidal zone routinely survive predation attempts that result in shell damage (up to 8% of *M. edulis* field populations, see Figure 1) and are capable of quickly closing and mineralizing repair patches despite variation in environmental pCO_2. Additionally, the decline in physiological condition of mussels within all laboratory treatments supports the conclusion that shell repair is energy-intensive. Given its importance to survival across a wide range of molluscan species, shell repair may impose energetic limitations on other physiological processes such as growth or reproduction [103]. Our results suggest that in areas where bivalves sustain a high rate of shell damage, it is possible that the costs associated with shell repair could compound over time, preventing smaller individuals from quickly surpassing the size range in which larger predators (e.g., sea stars, crustaceans) can handle them [104]. To tease apart these interactions, future work would benefit from integrating biomechanical, proteomic, and transcriptomic techniques to describe the shell repair process in different environmental conditions and under different degrees of food limitation.

Supplementary Materials: The following supporting information can be downloaded at: https://www.mdpi.com/article/10.3390/jmse10030359/s1; Figure S1: Mussel physiological and reproductive condition; Figure S2: Impact of mussel condition and reproductive status on shell repair; Table S1: Impact of pCO_2 on shell repair timing (ANCOVA); Table S2: Impact of pCO_2 and CI on repaired shell properties within endpoint populations (ANCOVA); Table S3: Impact of pCO_2 on a variety of physiological and shell parameters across all treatments and sampling timepoints (ANCOVA); Table S4: Impact of pCO_2 on the mean and max grayscale μCT grayscale values of endpoint samples (ANOVA); Table S5: Comparison of initial and final mussel physiological and reproductive condition (ANOVA).

Author Contributions: M.N.G., M.J.O. and E.C. conceived of the study. M.C. and E.C. conducted mussel bed surveys and performed the shell repair field assay using *M. edulis*. M.N.G. and M.J.O. conducted the ocean acidification experiment with *M. trossulus* and completed mechanical testing, material composition assays, and μCT analyses. M.N.G. conducted the SEM study. M.N.G. and E.C. performed data analysis and wrote the manuscript. All authors have read and agreed to the published version of the manuscript.

Funding: This work was supported by NSF award #DBI-0829486 to K. Sebens, T. Klinger, and J. Murray and by NSF awards #EF-104113 and OCE-2050273 to E. Carrington. Part of this work was conducted at the Molecular Analysis Facility, a National Nanotechnology Coordinated Infrastructure site at the University of Washington which is supported in part by the National Science Foundation award ECC-1542101, the University of Washington (UW), the Molecular Engineering and Sciences Institute, the Clean Energy Institute, and the National Institutes of Health. Additional support was provided by a UW Turn Point Faculty Fellowship to E. Carrington.

Data Availability Statement: Data are archived under project #2250 at www.bco-dmo.org (accessed on 2 January 2022).

Acknowledgments: We are grateful for the logistical and technical support of Rainah Sandstrom, Nicki LeBaron, Hilary Hayford, Laura Newcomb, and Michelle Herko. A special thank you to Timothy Cox at Seattle Children's Research Institute and Ben Marwick of the University of Washington Anthropology Department for sharing their time, expertise, and resources. We would also like to thank the efforts of Liz Harper and one anonymous reviewer for their insightful comments that greatly improved the quality of this manuscript.

Conflicts of Interest: The authors declare no conflict of interest.

References

1. Gutierrez, J.L.; Jones, C.G.; Strayer, D.L.; Iribarne, O.O. Mollusks as ecosystem engineers: The role of shell production in aquatic habitats. *Oikos* **2003**, *101*, 79–90. [CrossRef]
2. Vaughn, C.C.; Hoellein, T.J. Bivalve impacts in freshwater and marine ecosystems. *Annu. Rev. Ecol. Evol. Syst.* **2018**, *49*, 183–208. [CrossRef]
3. FAO. Fisheries and Aquaculture topics. The State of World Fisheries and Aquaculture (SOFIA). Topics Fact Sheets. Text by Jean-Francois Pulvenis. In *FAO Fisheries Division*; FAO: Rome, Italy, 2020.
4. Vermeij, G.J. *A Natural History of Shells*; Princeton University Press: Princeton, NJ, USA, 1995; Volume 15.
5. Hofmann, G.E.; Smith, J.E.; Johnson, K.S.; Send, U.; Levin, L.A.; Micheli, F.; Paytan, A.; Price, N.N.; Peterson, B.; Takeshita, Y. High-frequency dynamics of ocean pH: A multi-ecosystem comparison. *PLoS ONE* **2011**, *6*, e28983. [CrossRef]
6. Frieder, C.; Nam, S.; Martz, T.; Levin, L. High temporal and spatial variability of dissolved oxygen and pH in a nearshore California kelp forest. *Biogeosciences* **2012**, *9*, 3917–3930. [CrossRef]

7. George, M.N.; Andino, J.; Huie, J.; Carrington, E. Microscale pH and dissolved oxygen fluctuations within mussel aggregations and their implications for mussel attachment and raft aquaculture. *J. Shellfish Res.* **2019**, *38*, 795–809. [CrossRef]
8. Lowe, A.T.; Bos, J.; Ruesink, J. Ecosystem metabolism drives pH variability and modulates long-term ocean acidification in the Northeast Pacific coastal ocean. *Sci. Rep.* **2019**, *9*, 963. [CrossRef]
9. Bates, N.; Best, M.; Neely, K.; Garley, R.; Dickson, A.; Johnson, R. Detecting anthropogenic carbon dioxide uptake and ocean acidification in the North Atlantic Ocean. *Biogeosci. Discuss.* **2012**, *9*, 2509–2522. [CrossRef]
10. Doney, S.C.; Fabry, V.J.; Feely, R.A.; Kleypas, J.A. *Ocean Acidification: The Other CO_2 Problem*; Gulf Professional Publishing: Houston, TX, USA, 2009. [CrossRef]
11. Gazeau, F.; Parker, L.M.; Comeau, S.; Gattuso, J.-P.; O'Connor, W.A.; Martin, S.; Pörtner, H.-O.; Ross, P.M. Impacts of ocean acidification on marine shelled molluscs. *Mar. Biol.* **2013**, *160*, 2207–2245. [CrossRef]
12. Clements, J.C.; Chopin, T. Ocean acidification and marine aquaculture in North America: Potential impacts and mitigation strategies. *Rev. Aquac.* **2017**, *9*, 326–341. [CrossRef]
13. Halpern, B.S.; Walbridge, S.; Selkoe, K.A.; Kappel, C.V.; Micheli, F.; D'Agrosa, C.; Bruno, J.F.; Casey, K.S.; Ebert, C.; Fox, H.E. A global map of human impact on marine ecosystems. *Science* **2008**, *319*, 948–952. [CrossRef]
14. Carrington, E.; Moeser, G.M.; Dimond, J.; Mello, J.J.; Boller, M.L. Seasonal disturbance to mussel beds: Field test of a mechanistic model predicting wave dislodgment. *Limnol. Oceanogr.* **2009**, *54*, 978–986. [CrossRef]
15. Menge, B.A.; Berlow, E.L.; Blanchette, C.A.; Navarrete, S.A.; Yamada, S.B. The keystone species concept: Variation in interaction strength in a rocky intertidal habitat. *Ecol. Monogr.* **1994**, *64*, 249–286. [CrossRef]
16. Paine, R.T. Intertidal community structure. *Oecologia* **1974**, *15*, 93–120. [CrossRef]
17. Elner, R. The mechanics of predation by the shore crab, *Carcinus maenas* (L.), on the edible mussel, *Mytilus edulis* L. *Oecologia* **1978**, *36*, 333–344. [CrossRef]
18. Ebling, F.; Kitching, J.; Muntz, L.; Taylor, C.M. The ecology of Lough Ine. XIII Experimental observations of the destruction of *Mytilus edulis* and *Nucella lapillus* by crabs. *J. Anim. Ecol.* **1964**, *33*, 73–82. [CrossRef]
19. Clelland, E.S.; Saleuddin, A. Vacuolar-type ATPase in the accessory boring organ of *Nucella lamellosa* (Gmelin)(Mollusca: Gastropoda): Role in shell penetration. *Biol. Bull.* **2000**, *198*, 272–283. [CrossRef]
20. Norton-Griffiths, M. Some ecological aspects of the feeding behaviour of the oystercatcher *Haematopus ostralegus* on the edible mussel *Mytilus edulis*. *IBIS* **1967**, *109*, 412–424. [CrossRef]
21. Cintra-Buenrostro, C.E. Trampling, peeling and nibbling mussels: An experimental assessment of mechanical and predatory damage to shells of *Mytilus trossulus* (Mollusca: Mytilidae). *J. Shellfish Res.* **2007**, *26*, 221–231. [CrossRef]
22. Vermeij, G.J. Traces and trends of predation, with special reference to bivalved animals. *Palaeontology* **1983**, *26*, 455–465.
23. Vermeij, G.J. Unsuccessful predation and evolution. *Am. Nat.* **1982**, *120*, 701–720. [CrossRef]
24. Kaehler, S.; McQuaid, C. Lethal and sub-lethal effects of phototrophic endoliths attacking the shell of the intertidal mussel Perna perna. *Mar. Biol.* **1999**, *135*, 497–503. [CrossRef]
25. Christensen, A.M. Feeding biology of the sea star *Astropecten irregularis* Pennant. *Ophelia* **1970**, *8*, 1–134.
26. Carriker, M.R.; Zandt, D.V. Predatory behavior of a shell-boring muricid gastropod. In *Behavior of Marine Animals*; Springer: Berlin, Germany, 1972; pp. 157–244.
27. Palmer, A.R.; Szymanska, J.; Thomas, L. Prolonged withdrawal: A possible predator evasion behavior in *Balanus glandula* (Crustacea: Cirripedia). *Mar. Biol.* **1982**, *67*, 51–55. [CrossRef]
28. Xiong, X.; Cao, Y.; Li, Z.; Jiao, Y.; Du, X.; Zheng, Z. Transcriptome analysis reveals the transition and crosslinking of immune response and biomineralization in shell damage repair in pearl oyster. *Aquac. Rep.* **2021**, *21*, 100851. [CrossRef]
29. Burnett, L.E. Physiological responses to air exposure: Acid-base balance and the role of branchial water stores. *Am. Zool.* **1988**, *28*, 125–135. [CrossRef]
30. Meenakshi, V.; Blackwelder, P.L.; Wilbur, K.M. An ultrastructural study of shell regeneration in *Mytilus edulis* (Mollusca: Bivalvia). *J. Zool.* **1973**, *171*, 475–484. [CrossRef]
31. Sleight, V.A.; Thorne, M.A.; Peck, L.S.; Clark, M.S. Transcriptomic response to shell damage in the Antarctic clam, *Laternula elliptica*: Time scales and spatial localisation. *Mar. Genom.* **2015**, *20*, 45–55. [CrossRef]
32. Beedham, G. *Repair of the Shell in Species of Anodonta*; Zoological Society of London: London, UK, 1965; pp. 107–123.
33. Mount, A.S.; Wheeler, A.; Paradkar, R.P.; Snider, D. Hemocyte-mediated shell mineralization in the eastern oyster. *Science* **2004**, *304*, 297–300. [CrossRef]
34. Reed-Miller, C. The initial calcification process in shell-regenerating Tegula (Archaeogastropoda). *Biol. Bull.* **1983**, *165*, 265–275. [CrossRef]
35. LaBarbera, M.; Merz, R.A. Postmortem changes in strength of gastropod shells: Evolutionary implications for hermit crabs, snails, and their mutual predators. *Paleobiology* **1992**, *18*, 367–377. [CrossRef]
36. O'Neill, M.; Mala, R.; Cafiso, D.; Bignardi, C.; Taylor, D. Repair and remodelling in the shells of the limpet *Patella vulgata*. *J. R. Soc. Interface* **2018**, *15*, 20180299. [CrossRef] [PubMed]
37. Crane, R.; Diaz Reyes, J.; Denny, M. Bivalves rapidly repair shells damaged by fatigue and bolster strength. *J. Exp. Biol.* **2021**, *224*, jeb242681. [CrossRef] [PubMed]
38. Dunachie, J. XV—The periostracum of Mytilus edulis. *Earth Environ. Sci. Trans. R. Soc. Edinb.* **1963**, *65*, 383–411. [CrossRef]

39. Checa, A.G.; Pina, C.M.; Osuna-Mascaró, A.J.; Rodríguez-Navarro, A.B.; Harper, E.M. Crystalline organization of the fibrous prismatic calcitic layer of the Mediterranean mussel *Mytilus galloprovincialis*. *Eur. J. Mineral.* **2014**, *26*, 495–505. [CrossRef]
40. Bevelander, G.; Nakahara, H. An electron microscope study of the formation of the nacreous layer in the shell of certain bivalve molluscs. *Calcif. Tissue Res.* **1969**, *3*, 84–92. [CrossRef]
41. Kadar, E.; Lobo-da-Cunha, A.; Azevedo, C. Mantle-to-shell $CaCO_3$ transfer during shell repair at different hydrostatic pressures in the deep-sea vent mussel *Bathymodiolus azoricus* (Bivalvia: Mytilidae). *Mar. Biol.* **2009**, *156*, 959–967. [CrossRef]
42. Saleuddin, A.; Chan, W. Shell regeneration in Helix: Shell matrix composition and crystal formation. *Can. J. Zool.* **1969**, *47*, 1107–1111. [CrossRef]
43. Abolinš-Krogis, A. Ultrastructural study of the shell-repair membrane in the snail, *Helix pomatia* L. *Cell Tissue Res.* **1976**, *172*, 455–476. [CrossRef]
44. Yarra, T.; Blaxter, M.; Clark, M.S. A bivalve biomineralization toolbox. *Mol. Biol. Evol.* **2021**, *38*, 4043–4055. [CrossRef]
45. Yarra, T.; Ramesh, K.; Blaxter, M.; Hüning, A.; Melzner, F.; Clark, M.S. Transcriptomic analysis of shell repair and biomineralization in the blue mussel, *Mytilus edulis*. *BMC Genom.* **2021**, *22*, 7. [CrossRef]
46. Checa, A.G. Physical and biological determinants of the fabrication of molluscan shell microstructures. *Front. Mar. Sci.* **2018**, 353. [CrossRef]
47. Hüning, A.K.; Lange, S.M.; Ramesh, K.; Jacob, D.E.; Jackson, D.J.; Panknin, U.; Gutowska, M.A.; Philipp, E.E.; Rosenstiel, P.; Lucassen, M. A shell regeneration assay to identify biomineralization candidate genes in mytilid mussels. *Mar. Genom.* **2016**, *27*, 57–67. [CrossRef] [PubMed]
48. Orr, J.C.; Fabry, V.J.; Aumont, O.; Bopp, L.; Doney, S.C.; Feely, R.A.; Gnanadesikan, A.; Gruber, N.; Ishida, A.; Joos, F. Anthropogenic ocean acidification over the twenty-first century and its impact on calcifying organisms. *Nature* **2005**, *437*, 681–686. [CrossRef]
49. Ries, J.B.; Cohen, A.L.; McCorkle, D.C. Marine calcifiers exhibit mixed responses to CO_2-induced ocean acidification. *Geology* **2009**, *37*, 1131–1134. [CrossRef]
50. Zhao, X.; Han, Y.; Chen, B.; Xia, B.; Qu, K.; Liu, G. CO_2-driven ocean acidification weakens mussel shell defense capacity and induces global molecular compensatory responses. *Chemosphere* **2020**, *243*, 125415. [CrossRef] [PubMed]
51. Gazeau, F.; Gattuso, J.-P.; Dawber, C.; Pronker, A.; Peene, F.; Peene, J.; Heip, C.; Middelburg, J. Effect of ocean acidification on the early life stages of the blue mussel *Mytilus edulis*. *Biogeosciences* **2010**, *7*, 2051. [CrossRef]
52. Kurihara, H. Effects of CO_2-driven ocean acidification on the early developmental stages of invertebrates. *Mar. Ecol. Prog. Ser.* **2008**, *373*, 275–284. [CrossRef]
53. Gazeau, F.; Alliouane, S.; Bock, C.; Bramanti, L.; López Correa, M.; Gentile, M.; Hirse, T.; Pörtner, H.-O.; Ziveri, P. Impact of ocean acidification and warming on the Mediterranean mussel (*Mytilus galloprovincialis*). *Front. Mar. Sci.* **2014**, *1*, 62. [CrossRef]
54. Mackenzie, C.L.; Ormondroyd, G.A.; Curling, S.F.; Ball, R.J.; Whiteley, N.M.; Malham, S.K. Ocean warming, more than acidification, reduces shell strength in a commercial shellfish species during food limitation. *PLoS ONE* **2014**, *9*, e86764. [CrossRef]
55. Vihtakari, M.; Hendriks, I.E.; Holding, J.; Renaud, P.E.; Duarte, C.M.; Havenhand, J.N. Effects of ocean acidification and warming on sperm activity and early life stages of the Mediterranean mussel (*Mytilus galloprovincialis*). *Water* **2013**, *5*, 1890–1915. [CrossRef]
56. Waldbusser, G.G.; Hales, B.; Langdon, C.J.; Haley, B.A.; Schrader, P.; Brunner, E.L.; Gray, M.W.; Miller, C.A.; Gimenez, I. Saturation-state sensitivity of marine bivalve larvae to ocean acidification. *Nat. Clim. Chang.* **2015**, *5*, 273–280. [CrossRef]
57. Fitzer, S.C.; Phoenix, V.R.; Cusack, M.; Kamenos, N.A. Ocean acidification impacts mussel control on biomineralisation. *Sci. Rep.* **2014**, *4*, 6218. [CrossRef] [PubMed]
58. Fitzer, S.C.; Vittert, L.; Bowman, A.; Kamenos, N.A.; Phoenix, V.R.; Cusack, M. Ocean acidification and temperature increase impact mussel shell shape and thickness: Problematic for protection? *Ecol. Evol.* **2015**, *5*, 4875–4884. [CrossRef] [PubMed]
59. Palmer, A. Relative cost of producing skeletal organic matrix versus calcification: Evidence from marine gastropods. *Mar. Biol.* **1983**, *75*, 287–292. [CrossRef]
60. Palmer, A.R. Calcification in marine molluscs: How costly is it? *Proc. Natl. Acad. Sci. USA* **1992**, *89*, 1379–1382. [CrossRef]
61. Gazeau, F.; Quiblier, C.; Jansen, J.M.; Gattuso, J.P.; Middelburg, J.J.; Heip, C.H. Impact of elevated CO_2 on shellfish calcification. *Geophys. Res. Lett.* **2007**, *34*. [CrossRef]
62. Pörtner, H.O.; Langenbuch, M.; Reipschläger, A. Biological impact of elevated ocean CO_2 concentrations: Lessons from animal physiology and earth history. *J. Oceanogr.* **2004**, *60*, 705–718. [CrossRef]
63. Waldbusser, G.G.; Hales, B.; Haley, B.A. Calcium carbonate saturation state: On myths and this or that stories. *ICES J. Mar. Sci.* **2016**, *73*, 563–568. [CrossRef]
64. Carrington, E. Seasonal variation in the attachment strength of blue mussels: Causes and consequences. *Limnol. Oceanogr.* **2002**, *47*, 1723–1733. [CrossRef]
65. Carrington, E. The ecomechanics of mussel attachment: From molecules to ecosystems. *Integr. Comp. Biol.* **2002**, *42*, 846–852. [CrossRef]
66. Hilbish, T.J.; Bayne, B.L.; Day, A. Genetics of physiological differentiation within the marine mussel genus *Mytilus*. *Evolution* **1994**, *48*, 267–286. [CrossRef]
67. Varvio, S.-L.; Koehn, R.K.; Väinölä, R. Evolutionary genetics of the *Mytilus edulis* complex in the North Atlantic region. *Mar. Biol.* **1988**, *98*, 51–60. [CrossRef]

68. Saavedra, C.; Stewart, D.T.; Stanwood, R.R.; Zouros, E. Species-specific segregation of gender-associated mitochondrial DNA types in an area where two mussel species (*Mytilus edulis* and *M. trossulus*) hybridize. *Genetics* **1996**, *143*, 1359–1367. [CrossRef] [PubMed]
69. Comesaña, A.; Toro, J.; Innes, D.; Thompson, R. A molecular approach to the ecology of a mussel (*Mytilus edulis–M. trossulus*) hybrid zone on the east coast of Newfoundland, Canada. *Mar. Biol.* **1999**, *133*, 213–221. [CrossRef]
70. Toro, J.; Innes, D.; Thompson, R. Genetic variation among life-history stages of mussels in a *Mytilus edulis–M. trossulus* hybrid zone. *Mar. Biol.* **2004**, *145*, 713–725. [CrossRef]
71. Penney, R.W.; Hart, M.J.; Templeman, N.D. Shell strength and appearance in cultured blue mussels *Mytilus edulis, M. trossulus,* and *M. edulis* × *M. trossulus* hybrids. *N. Am. J. Aquac.* **2007**, *69*, 281–295. [CrossRef]
72. Carboni, S.; Evans, S.; Tanner, K.E.; Davie, A.; Bekaert, M.; Fitzer, S.C. Are Shell Strength Phenotypic Traits in Mussels Associated with Species Alone? *Aquac. J.* **2021**, *1*, 3–13. [CrossRef]
73. Baird, R. Measurement of condition in mussels and oysters. *ICES J. Mar. Sci.* **1958**, *23*, 249–257. [CrossRef]
74. Rueden, C.T.; Schindelin, J.; Hiner, M.C.; DeZonia, B.E.; Walter, A.E.; Arena, E.T.; Eliceiri, K.W. ImageJ2: ImageJ for the next generation of scientific image data. *BMC Bioinform.* **2017**, *18*, 529. [CrossRef]
75. Bell, E.; Gosline, J. Mechanical design of mussel byssus: Material yield enhances attachment strength. *J. Exp. Biol.* **1996**, *199*, 1005–1017. [CrossRef]
76. Palmerini, P.; Bianchi, C. Biomass measurements and weight-to-weight conversion factors: A comparison of methods applied to the mussel *Mytilus galloprovincialis*. *Mar. Biol.* **1994**, *120*, 273–277. [CrossRef]
77. O'Donnell, M.J.; George, M.N.; Carrington, E. Mussel byssus attachment weakened by ocean acidification. *Nat. Clim. Chang.* **2013**, *3*, 587–590. [CrossRef]
78. Martz, T.R.; Connery, J.G.; Johnson, K.S. Testing the Honeywell Durafet® for seawater pH applications. *Limnol. Oceanogr. Methods* **2010**, *8*, 172–184. [CrossRef]
79. Robbins, L.; Hansen, M.; Kleypas, J.; Meylan, S. CO_2calc—A user-friendly seawater carbon calculator for Windows, Max OS X, and iOS (iPhone). *US Geol. Surv. Open-File Rep.* **2010**, *1280*, 2010.
80. Orr, J.C.; Epitalon, J.-M.; Dickson, A.G.; Gattuso, J.-P. Routine uncertainty propagation for the marine carbon dioxide system. *Mar. Chem.* **2018**, *207*, 84–107. [CrossRef]
81. Hu, Y.; Limaye, A.; Lu, J. A new tool for 3D segmentation of computed tomography data: Drishti Paint and its applications. *bioRxiv* **2020**. [CrossRef]
82. Fernandez, E. *Johnson: Johnson Transformation*; R Package Version 1.4.; Johnson & Johnson: New Brunswick, NJ, USA, 2014.
83. de Mendiburu, M. Agricolae: Statistical procedures for agricultural research. *Am. J. Plant Sci.* **2017**, *8*, 7.
84. Liao, Z.; Bao, L.-F.; Fan, M.-H.; Gao, P.; Wang, X.-X.; Qin, C.-L.; Li, X.-M. In-depth proteomic analysis of nacre, prism, and myostracum of *Mytilus shell*. *J. Proteom.* **2015**, *122*, 26–40. [CrossRef]
85. Marie, B.; Le Roy, N.; Zanella-Cléon, I.; Becchi, M.; Marin, F. Molecular evolution of mollusc shell proteins: Insights from proteomic analysis of the edible mussel Mytilus. *J. Mol. Evol.* **2011**, *72*, 531–546. [CrossRef]
86. Gao, P.; Liao, Z.; Wang, X.-X.; Bao, L.-F.; Fan, M.-H.; Li, X.-M.; Wu, C.-W.; Xia, S.-W. Layer-by-layer proteomic analysis of *Mytilus galloprovincialis* shell. *PLoS ONE* **2015**, *10*, e0133913. [CrossRef]
87. Yin, Y.; Huang, J.; Paine, M.L.; Reinhold, V.N.; Chasteen, N.D. Structural characterization of the major extrapallial fluid protein of the mollusc *Mytilus edulis*: Implications for function. *Biochemistry* **2005**, *44*, 10720–10731. [CrossRef] [PubMed]
88. Hattan, S.J.; Laue, T.M.; Chasteen, N.D. Purification and characterization of a novel calcium-binding protein from the extrapallial fluid of the mollusc, *Mytilus edulis*. *J. Biol. Chem.* **2001**, *276*, 4461–4468. [CrossRef] [PubMed]
89. Wheeler, A.; George, J.W.; Evans, C. Control of calcium carbonate nucleation and crystal growth by soluble matrx of oyster shell. *Science* **1981**, *212*, 1397–1398. [CrossRef] [PubMed]
90. Falini, G.; Albeck, S.; Weiner, S.; Addadi, L. Control of aragonite or calcite polymorphism by mollusk shell macromolecules. *Science* **1996**, *271*, 67–69. [CrossRef]
91. Marin, F.; Luquet, G.; Marie, B.; Medakovic, D. Molluscan shell proteins: Primary structure, origin, and evolution. *Curr. Top. Dev. Biol.* **2007**, *80*, 209–276.
92. Weiss, I.M.; Tuross, N.; Addadi, L.; Weiner, S. Mollusc larval shell formation: Amorphous calcium carbonate is a precursor phase for aragonite. *J. Exp. Zool.* **2002**, *293*, 478–491. [CrossRef]
93. Blundon, J.; Vermeij, G. Effect of shell repair on shell strength in the gastropod *Littorina irrorata*. *Mar. Biol.* **1983**, *76*, 41–45. [CrossRef]
94. Uozumi, S.; Suzuki, S. "Organic Membrane-Shell" and Initial Calcification in Shell Regeneration. *J. Fac. Sci. Hokkaido Univ. Ser. IV* **1979**, *19*, 37–74.
95. Li, S.; Liu, C.; Huang, J.; Liu, Y.; Zheng, G.; Xie, L.; Zhang, R. Interactive effects of seawater acidification and elevated temperature on biomineralization and amino acid metabolism in the mussel *Mytilus edulis*. *J. Exp. Biol.* **2015**, *218*, 3623–3631. [CrossRef]
96. Hahn, S.; Rodolfo-Metalpa, R.; Griesshaber, E.; Schmahl, W.W.; Buhl, D.; Hall-Spencer, J.; Baggini, C.; Fehr, K.; Immenhauser, A. Marine bivalve shell geochemistry and ultrastructure from modern low pH environments: Environmental effect versus experimental bias. *Biogeosciences* **2012**, *9*, 1897–1914. [CrossRef]
97. Gaylord, B.; Hill, T.M.; Sanford, E.; Lenz, E.A.; Jacobs, L.A.; Sato, K.N.; Russell, A.D.; Hettinger, A. Functional impacts of ocean acidification in an ecologically critical foundation species. *J. Exp. Biol.* **2011**, *214*, 2586–2594. [CrossRef] [PubMed]

98. Melzner, F.; Stange, P.; Trübenbach, K.; Thomsen, J.; Casties, I.; Panknin, U.; Gorb, S.N.; Gutowska, M.A. Food supply and seawater pCO_2 impact calcification and internal shell dissolution in the blue mussel *Mytilus edulis*. *PLoS ONE* **2011**, *6*, e24223. [CrossRef] [PubMed]
99. Wolowicz, M.; Goulletquer, P. The shell organic content in the energy budget of *Mytilus trossulus* from the South Baltic. *Haliotis* **1999**, *28*, 1–10.
100. Melzner, F.; Thomsen, J.; Koeve, W.; Oschlies, A.; Gutowska, M.A.; Bange, H.W.; Hansen, H.P.; Körtzinger, A. Future ocean acidification will be amplified by hypoxia in coastal habitats. *Mar. Biol.* **2013**, *160*, 1875–1888. [CrossRef]
101. Thomsen, J.; Casties, I.; Pansch, C.; Körtzinger, A.; Melzner, F. Food availability outweighs ocean acidification effects in juvenile *Mytilus edulis*: Laboratory and field experiments. *Glob. Chang. Biol.* **2013**, *19*, 1017–1027. [CrossRef]
102. Hüning, A.K.; Melzner, F.; Thomsen, J.; Gutowska, M.A.; Krämer, L.; Frickenhaus, S.; Rosenstiel, P.; Pörtner, H.-O.; Philipp, E.E.; Lucassen, M. Impacts of seawater acidification on mantle gene expression patterns of the Baltic Sea blue mussel: Implications for shell formation and energy metabolism. *Mar. Biol.* **2013**, *160*, 1845–1861. [CrossRef]
103. Roberts, E.A.; Newcomb, L.A.; McCartha, M.M.; Harrington, K.J.; LaFramboise, S.A.; Carrington, E.; Sebens, K.P. Resource allocation to a structural biomaterial: Induced production of byssal threads decreases growth of a marine mussel. *Funct. Ecol.* **2021**, *35*, 1222–1239. [CrossRef]
104. Paine, R.T. Size-limited predation: An observational and experimental approach with the *Mytilus-Pisaster* interaction. *Ecology* **1976**, *57*, 858–873. [CrossRef]

Journal of
*Marine Science
and Engineering*

MDPI

Article

Artificial Intelligence as a Tool to Study the 3D Skeletal Architecture in Newly Settled Coral Recruits: Insights into the Effects of Ocean Acidification on Coral Biomineralization

Federica Scucchia [1,2,*], Katrein Sauer [3], Paul Zaslansky [3,*,†] and Tali Mass [1,4,†]

1 Department of Marine Biology, Leon H. Charney School of Marine Sciences, University of Haifa, Haifa 3498838, Israel; tmass@univ.haifa.ac.il
2 The Interuniversity Institute of Marine Sciences, Eilat 88103, Israel
3 Department for Operative and Preventive Dentistry, Charité-Universitätsmedizin, 14497 Berlin, Germany; katrein.sauer@gmail.com
4 The Morris Kahn Marine Research Station, Sdot Yam, Haifa 30889, Israel
* Correspondence: federica.scucchia@studio.unibo.it (F.S.); paul.zaslansky@charite.de (P.Z.)
† These authors contributed equally to this work.

Abstract: Understanding the formation of the coral skeleton has been a common subject uniting various marine and materials study fields. Two main regions dominate coral skeleton growth: Rapid Accretion Deposits (RADs) and Thickening Deposits (TDs). These have been extensively characterized at the 2D level, but their 3D characteristics are still poorly described. Here, we present an innovative approach to combine synchrotron phase contrast-enhanced microCT (PCE-CT) with artificial intelligence (AI) to explore the 3D architecture of RADs and TDs within the coral skeleton. As a reference study system, we used recruits of the stony coral *Stylophora pistillata* from the Red Sea, grown under both natural and simulated ocean acidification conditions. We thus studied the recruit's skeleton under both regular and morphologically-altered acidic conditions. By imaging the corals with PCE-CT, we revealed the interwoven morphologies of RADs and TDs. Deep-learning neural networks were invoked to explore AI segmentation of these regions, to overcome limitations of common segmentation techniques. This analysis yielded highly-detailed 3D information about the RAD's and TD's architecture. Our results demonstrate how AI can be used as a powerful tool to obtain 3D data essential for studying coral biomineralization and for exploring the effects of environmental change on coral growth.

Keywords: coral reefs; coral recruits; biomineralization; skeletal structure; synchrotron phase contrast-enhanced microCT; PCE-CT; artificial intelligence; ocean acidification

Citation: Scucchia, F.; Sauer, K.; Zaslansky, P.; Mass, T. Artificial Intelligence as a Tool to Study the 3D Skeletal Architecture in Newly Settled Coral Recruits: Insights into the Effects of Ocean Acidification on Coral Biomineralization. *J. Mar. Sci. Eng.* **2022**, *10*, 391. https://doi.org/10.3390/jmse10030391

Academic Editors: Hildegard Westphal, Justin Ries and Steve Doo

Received: 13 January 2022
Accepted: 5 March 2022
Published: 9 March 2022

1. Introduction

In tropical and subtropical oceans, symbiotic stony corals provide an ecological framework that retains nutrients, support high rates of autotrophic production, and harbor extensive biological diversity. The skeletons, created by tiny coral polyps in colonies, are known to be among the largest bioconstructions in the world. Understanding the details of structure generation and principles of coral skeletal formation have been a common topic among diverse fields of study including biology, geochemistry, geology, paleontology, and materials science. The skeletons of stony corals consist of two main skeletal regions known as Rapid Accretion Deposits (RADs; also known as "calcification centers") and Thickening Deposits (TDs; also called "fibers"). RADs correspond to areas of the skeleton with rapid calcium carbonate deposition where new skeletal growth zones start to form [1–3]. They contain an organic matrix and consist of granular aggregates that comprise randomly oriented calcium carbonate crystals containing both aragonite and stable amorphous calcium carbonate [4–6]. A second calcium carbonate form, Mg-calcite, has been shown to also be present in the early stages of RADs formation, typical of initial mineral development in early coral life stages [7]. Outward from RADs, the mineral grows into TDs, forming elongated aragonite crystals with a fibrous morphology appearing compact and dense with significantly reduced amounts of organics as compared to RADs [2,3,8–11].

The coral skeleton has been well characterized at the micro-structural level [2–5,8–13], but less is known about how changes in the environment may affect its development, in particular regarding the formation of RADs and TDs. Recent predictions suggest that future ocean acidification (OA) conditions, resulting in a chemical shift toward higher seawater pCO_2 and lower pH, will significantly hinder the coral mineral formation and overall skeletal growth [14–18]. In the early stages of coral life, we have recently shown that exposure to future OA conditions causes a reduction in the abundance of RADs in the skeletal spines [19]. However, previous results relied solely on scanning electron microscope (SEM) images; hence, they were limited only to two-dimensional (2D) surface observations. Indeed, most studies on coral skeletal structures make use of 2D-based microscopy analyses [1–3,8,9,11–13,20], which lack volumetric quantifications. Therefore, the lack of an appropriate methodology to precisely visualize the volumetric configuration of RADs and TDs has limited investigation of the 3D structure of these two skeletal regions [3,11,21].

Following the widespread development of high-resolution X-ray μCT in medical and industrial applications, non-destructive imaging of 3D structures has become popular. It is thus possible to monitor mineral internal structural changes [22] and characterization [23] at resolutions down to a few tens of nanometers. Conventional X-ray μCT imaging methods mainly rely on X-ray absorption as a mechanism to generate images based on the attenuation of the X-rays by the material, yielding spatial information about density distributions within the sample. This technique, however, is prone to normalization artifacts and does not distinguish between materials with very similar X-ray attenuation [24,25]. Other methods such as phase contrast-enhanced microCT (PCE-CT) exploit differences in X-ray interaction to create high-contrast images where even small differences in density along the X-ray path can lead to strong contrast enhancement at interfaces between material phases [24]. PCE-CT imaging can thus highlight edges and internal boundaries in intact samples, even in the case of quite similar material compositions [24]. These characteristics of PCE-CT make it a good candidate for enhancing the visibility of RADs and TDs and ease their identification within the coral skeleton.

For a detailed analysis of microstructures, volumetric quantification techniques must be applied with the capacity to accurately differentiate and represent individual material phases. In recent years, several works have shown that artificial intelligence (AI) algorithms can be applied for visual classification tasks such as in the case of materials with complex microstructures [26–28]. These applications make increasing use of machine learning methods based on Deep Learning (DL) neural networks. DL networks comprise a large number of interconnected processors (neurons) that work in parallel [29]. For visual classification, the network is trained to identify structures of interest using a training "ground truth" dataset, which is used by the computer as a reference to classify similar images. The process is known as supervised classification, and has been employed in μCT images of minerals to visualize and characterize 3D structures with high accuracy [30–33].

In the present study, we combine AI and PCE-CT to obtain quantitative information about the density and 3D distribution of RADs and TDs within coral skeletons imaged with phase contrast enhancement. Such information is currently unavailable using conventional 2D or other imaging methods. We used coral primary polyps as the study system for the development of a robust analytic 3D quantification methodology, since rapid calcification during initial life stages of stony corals provides a unique opportunity to study the formation of calcium carbonate that involves extensive skeleton morphological changes. In addition, the early stages of coral development represent a good candidate to test the performance of our approach under experimental conditions that we know induce alterations in the development of RADs, such as exposure to OA conditions [19].

To develop our new methodology and optimize its performance across varied conditions, we analyzed high resolution scans of the skeletal structures of recruits of an abundant stony coral species *Stylophora pistillata* from the Red Sea, following settlement and growth under both natural and OA conditions. In particular, we simulated the global mean surface-ocean decline in pH predicted to occur under the RCP8.5 scenario [34] by the end of this century (pH drop ca. 0.1–0.4 units, from ~8.0 to 7.6; [35]). Employing PCE-CT, we identified the μm-sized edges of RADs and TDs within the skeleton. The high-contrast images generated with PCE-CT allowed us to use AI to segment RADs and TDs, under both natural conditions and under what we expect to be morphologically-altered OA conditions, and to reproduce these structures in 3D with quantitative detail.

2. Materials and Methods

2.1. Sample Collection and OA Experiment

Coral larvae were collected from 15 randomly selected adult colonies of the stony coral *S. pistillata* on the reef adjacent to the Interuniversity Institute of Marine Sciences (IUI, 29°30′06.0″ N 34°54′58.3″ E) in the Gulf of Eilat (Israel), under a permit from the Israeli Natural Parks Authority. Larvae were collected using dedicated traps (160 μm plankton net with a plastic collection container) that were placed on adult colonies from the shallow reef (depth 6–7 m) for several nights of spawning during April 2020.

All actively swimming larvae were pooled together and were transported to a controlled environment aquarium system at the Leon H. Charney School of Marine Science (University of Haifa, Haifa, Israel). Larvae were placed in custom-made polypropylene plastic chambers (~20 larvae per chamber) consisting of a central cylinder sealed at the ends by plankton netting. Settlement chambers were used in this experiment to avoid losing larvae through the water recycle system of the aquariums before settlement, as previously reported [19,36,37]. Prior to the insertion of the larvae, the chambers were washed to remove any potential chemical released by the plastic, and they were left to soak in seawater for one week.

The settlement chambers (1 chamber per aquarium) were placed in a system of 6 flow-through aquariums with artificial seawater (Red Sea Salt, Red Sea Ltd., Houston, TX, USA) replicating the spring northern Red Sea water conditions as previously described [19]: pH of 8.19 ± 0.01 (seasonal mean ± s.d.), salinity of 40.63 ± 0.03 g L^{-1} (seasonal mean ± s.d.), and temperature of 22.18 ± 0.17 °C (seasonal mean ± s.d.). An irradiance of 250 μmol photons m^{-2} s^{-1} on a 12-h light-dark cycle was provided by a Mitras LX 7206 LED aquarium light system (GHL, Kaiserslautern, Germany).

Prior to the insertion of the settlement chambers, the carbonate chemistry of seawater was manipulated in 3 of the experimental aquariums by injecting CO_2 to reduce the ambient pH 8.2 (pCO_2~487 μatm) and obtain the target value of pH 7.6 (pCO_2~1938 μatm). Temperature and salinity were monitored continuously throughout the experiment by electrodes linked to a monitoring system (GHL, Kaiserslautern, Germany) that controlled CO_2 injection, and they were also measured three times per day using a portable Orion Star™ A222 conductivity meter (Thermo Fisher Scientific, Waltham, MA, USA). Measurements of pH (NBS scale) were performed three times per day using a pH glass electrode (Metrohm, Herisau, Switzerland) where buffer solutions (Rocker Scientific, Taiwan) were employed to perform calibration (pH 4, pH 7, and pH 10). Measurements of total alkalinity (TA) were made once per day (triplicates) via titration with 0.1 N HCl containing 40.7 g NaCl L^{-1}, using an automatic alkalinity titrator (855 Robotic Titrosampler, Metrohm, Herisau, Switzerland) controlled by Tiamo (Software version 2.0, Metrohm, Herisau, Switzerland). Automated titrations of 50 mL samples were performed. Parameters of seawater carbonate system were calculated from pH, TA, temperature, and salinity using the CO2SYS package [38] with constants from [39] as refit by [40] (Supplemental Table S1). Throughout the experiment, corals were fed once per day with 2 mL of concentrated planktonic suspension (Microvore, Brightwell R aquatics, Fort Payne, AL, USA). The experiment lasted for a total of 9 days, following our previous experimental procedure [19], after which primary polyps were gently removed from the chambers and stored with 90% ethanol in 2 mL tubes.

2.2. X-ray μCT: Image Acquisition and Tomographic Reconstruction

Primary polyps were fixed with EpoFix resin (Agar Scientific, Stansted, UK) on top of polypropylene micropipette tips. Tomographic scanning was conducted at BAMline [41,42], the imaging beamline of BESSY II (the synchrotron storage ring of HZB—Helmholtz-Zentrum Berlin, Germany). Each sample (N = 3 per pH treatment) was attached to a metal stub and scanned with incremental rotation (multiple projections spanning 180°) (Figure 1a) using the high-resolution imaging setup (Supplemental Figure S1) [43] with exposure times set to 1 s. Projection images were acquired with a final pixel size of 2.2 μm using PCE-CT imaging mode at an energy of 24.5 keV. Additional absorption scans were performed for comparison, using an energy of 16.26 keV, to ensure correct identification of the skeletal structures avoiding contrast enhancement.

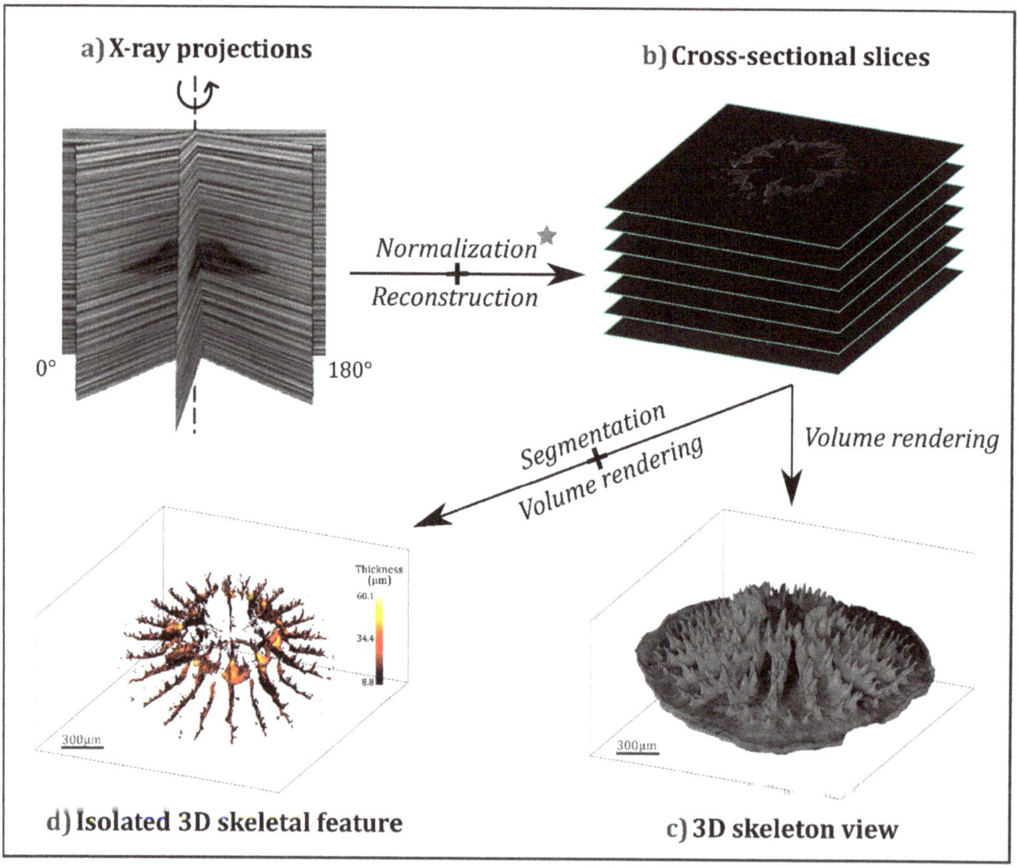

Figure 1. Workflow for acquisition and processing of tomographic datasets of coral recruit skeletons. (**a**) X-ray projections are acquired from different angles of the skeleton with incremental rotation (0° to 180°); (**b**) following normalization (gray star = additional details are shown in Figure 2), tomographic cross-sectional slices are reconstructed; (**c**) cross-sectional slices are stacked and processed into 3D views of the recruit skeletons; (**d**) AI-based segmentation and classification are performed so that the features of interest can be analyzed and visualized in 3D. An example distribution of RAD thicknesses is shown.

Prior to reconstruction, data were normalized to account for beam inhomogeneities using an in-house Octave-based reconstruction pipeline in the labs of the Charité, Universitätsmedizin (Berlin, Germany). Specifically, for each scan, the radiograms were background-corrected by normalization with best-fitting (minimum variance) empty beam (flat-field) images, obtained both before and after each scan, and corrected by subtraction of dark-current images (Figure 2). Reconstruction was performed by the filtered back projection method using nRecon (v 1.7.4.2, Brucker micro-CT, Kontich, Belgium) to generate 3D datasets from cross-sectional 2D images (Figure 1b). Tomographic datasets were visualized and further processed in 3D using Dragonfly (v 2021.3, Object Research Systems—ORS, Montreal, Quebec, Canada) (Figure 1c,d) [44], with additional stack analyses performed in FIJI [45].

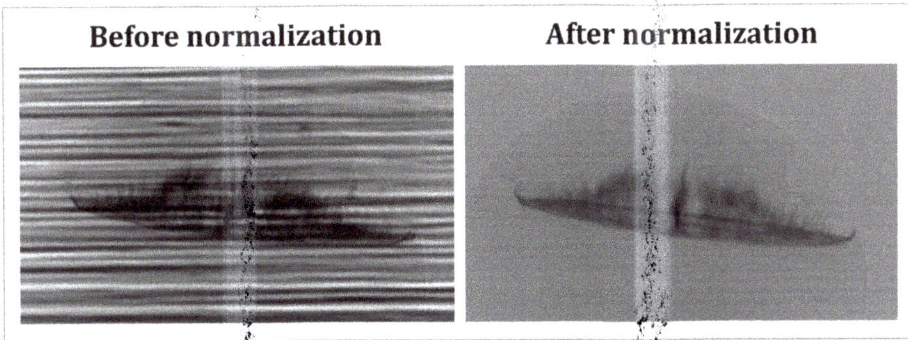

Figure 2. Example X-ray µCT radiographs of a recruit skeleton as obtained on the beamline (**left**) and after normalization (**right**). Radiographs were background-corrected by normalization with empty beam (flat-field) images and by the subtraction of dark-current images. The dark silhouette of the coral is seen to be suspended within a faint halo of epoxy resin, visible in the normalized (**right**) image.

2.3. Deep-Learning-Based Image Segmentation

To objectively identify structures in the PCE tomographic data, each reconstruction was segmented using the Deep Learning functions built into the 3D commercially-available software Dragonfly (v2021.3, Object Research Systems—ORS, Montreal, Quebec, Canada) [46]. Segmentation is necessary for the processes of analysis and quantification of the volume of RADs and TDs with respect to the total volume of the skeleton (mineral excluding air cavities) of each primary coral polyp scanned. This process is required to partition digital images into several sub-volumes, each representing a distinct 3D feature, obtained by identification and grouping of similar volume pixels (voxels) [47]. Image segmentation is thus typically used to locate objects of interest in large volumes of data needing quantification. In the present study, we applied semantic image segmentation by labeling and grouping pixels in each slice-image according to the feature that they were part of (RADs, TDs, or air spaces). For this classification, we used the U-Net network architecture algorithm, which has been shown to be very efficient in segmenting biomedical images [48]. When applying supervised classification, the network is manually trained by providing examples of how different features within the data should be classified. This is achieved by manually selecting groups of voxels and assigning them to a specific feature within a set of representative images, considered to be a training dataset. The computer then uses the manually-segmented training datasets as reference for automatic classification of all other regions in all images. We therefore manually delineated RADs, TDs, and air spaces in several (<10) 2D slices within each reconstructed tomographic dataset. These were used to train the network to recognize different features within all other images (Figure 3a–d).

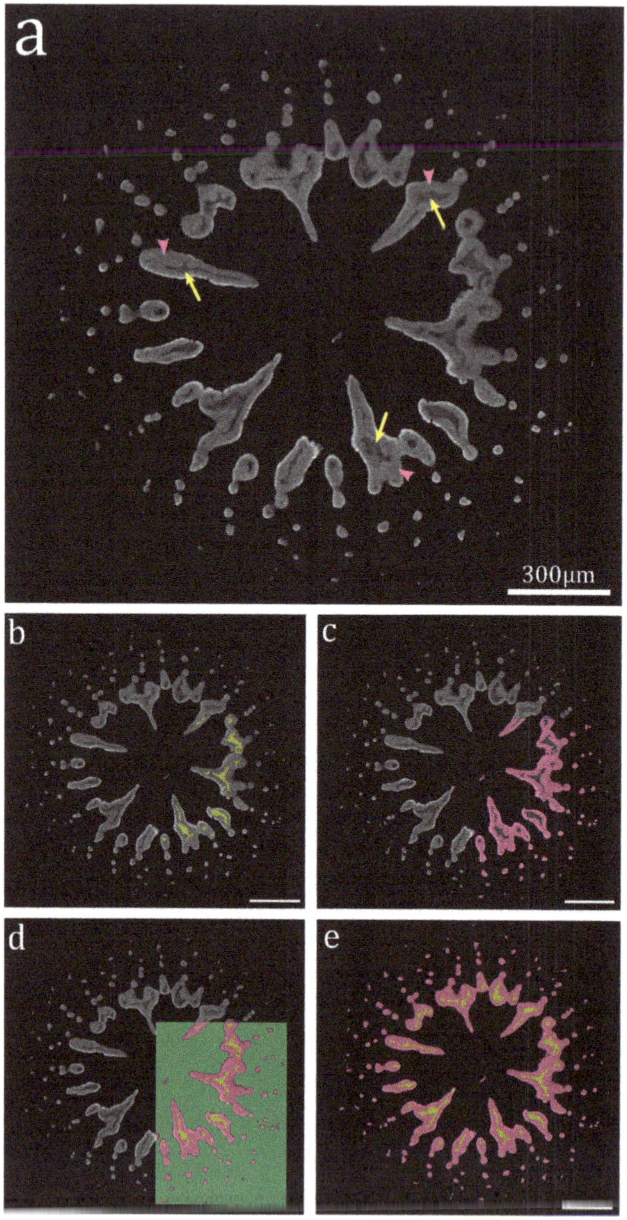

Figure 3. Overview of the DL (deep-learning-based) image segmentation process. (**a**) Example tomographic cross-sectional slices across the skeleton of a primary polyp before segmentation, revealing RADs (dark gray areas, yellow arrows) enclosed within TDs (light gray areas, pink arrowheads). (**b**,**c**) Manually delineated RADs (**b**; marked in yellow) and TDs (**c**; marked in pink) defined in a confined region within the training dataset. (**d**) Complete region manually delineated as training dataset, showing all three features (RADs in yellow, TDs in pink, and air spaces in green). (**e**) Result of the complete dataset processed by automatic DL-based segmentation. The AI network was applied to all slices within the stack, and it has automatically successfully distinguished RADs (yellow) and TDs (pink). Scale bars: 300 µm.

Training requires optimizing a series of parameters that define the performance of the U-Net network workflow. Within the Dragonfly platform, several parameters can be tuned: (1) patch size (images are split into smaller rectangular patches that contain the feature of interest [49]), (2) stride-to-input ratio (a parameter defining the typical position relationships between neighboring patches), (3) batch size (the number of patches evaluated concomitantly), (4) number of epochs (cycles of iteration that involve computation of all batches of the training dataset), (5) loss function (which measures the difference between groups of voxels in the training set images, considered to be "the ground truth images", and the output value predicted by the network [50]), (6) optimization algorithm (different approaches to minimize the loss function, and thus to reduce the training error), and (7) data augmentation (how many times each data patch is augmented during a single training epoch). In this work, optimization of the DL network performance was achieved by making trial runs with different combinations of number of training slices and of the above-mentioned values of network parameters (starting with default settings provided at first by the software), until reaching reproducible accuracy of the segmentation (a single user ensured by visual examination that the trained network accurately segmented skeletal regions, by comparison to the ground truth images, i.e., the training set of images). The final optimized workflow thus required training datasets of 8 to 11 slices and a series of parameters (1–6 as described above) set to: (1) 32, (2) 1, (3) 32, (4) 100, (5) OrsDiceLoss, (6) Adadelta, and (7) 2. The trained network was then applied to segment all slices within each tomographic stack (Figure 3e). The segmented skeletal structures were visualized in 3D (Figure 1d) and their thickness was computed.

2.4. *Performance of the DL-Based Image Segmentation and Evaluation of the RADs/TDs Shape Variability within and between Tomographic Datasets*

Each tomographic dataset (corresponding to a single coral recruit) was used to train a DL network independently. Representative slices from each dataset were trained to identify the 3 image features, as shown in Figure 3. Thereafter, networks optimized for each dataset were checked for segmentation of other datasets (e.g., DL network optimized on dataset A applied to segment dataset B). The degree to which image features were correctly classified as TDs or RADs was assessed by visual inspection and by computing the degree of mismatch (underestimation or overestimation in % of the RADs area with respect to the total area of the skeleton - measured as the combined cross-sectional RAD + TD areas in each slice). Performance was also compared with a DL network trained using all tomographic datasets (3 from the control pH and 3 from the low pH, 8–11 training slices were selected within each dataset) and using it to segment datasets from both the control and low pH conditions.

Because the performance of AI-based image segmentation is strongly influenced by the degree of variability of the target-objects shapes [51–54], we estimated the variation in the RADs-TDs relative shape configuration by using basic shape descriptors (perimeter and area per slice) and comparing them across datasets. In particular, we computed the RADs-TDs relative shape configuration according to Equation (1).

$$C = \frac{\frac{P(r)}{A(r)}}{P(t)/A(t)} \quad (1)$$

where $P(r)$ and $A(r)$ are the perimeter and area of RADs, respectively, and $P(t)$ and $A(t)$ are the perimeter and area of TDs, respectively. Such values were computed per each slice within each single dataset and ultimately expressed as a percentage. We then calculated the skewness of C distributions and used it as a metric to evaluate the degree of the RADs-TDs shape variation across datasets, as well as to compare it with the results of the segmentation performance.

2.5. Statistical Analysis

The per-slice percentage of RADs area over the cross-sectional area of the skeleton computed for different segmentation outcomes, as well as the skewness of the distribution of the per-slice RADs-TDs configuration, were tested for normality (Shapiro–Wilk test) and homogeneity of variance (Brown–Forsythe test). For the skewness measurements, the unpaired *t*-test was used (to test the differences between pH treatments), in which significant groups have a value of $p \leq 0.05$ (n = 3 primary polyps per pH condition). Nonparametric equivalents of tests were used in cases where the assumptions of normality and homogeneity of variance were violated. For such cases (i.e., mean per-slice percentage of RADs area over the total area of the skeleton), a Wilcoxon matched-pairs signed rank test was used, in which significant groups have a value of $p \leq 0.05$. The GraphPad Prism software version 9.0.0 (GraphPad Inc., San Diego, CA, USA) was used to perform the statistical tests. All results are presented as mean $\pm$ SE.

3. Results

3.1. Detection and Distinction of Coral Skeletal Structures

We reconstructed high-resolution tomography datasets of up to 1000 cross-sectional slices per scanned coral primary polyp, each amounting to a total of 54 GB of data. Each reconstructed image presents the complete form of the original coral skeleton including details of the internal skeletal architecture. In the PCE-CT images, strong differences in contrasts within the skeleton revealed clear internal variations in mineral distribution (Figure 4). Such variations coincide with previous observations of RADs possessing a less compact arrangement of the aragonite aggregates and a higher amount of organics compared to TDs [2,3,8–13]. Absorption images further confirm variations in mineral density between skeletal regions, though with blurry edges due to the poor signal. These scans show that RADs are indeed less dense than the surrounding TDs (~20% lower density; Supplemental Figure S2). This clear difference in density allowed us to ascertain that the contrasts observed by PCE-CT correspond to RADs and TDs in all phase contrast enhanced datasets. To allow for volumetric quantification and 3D visualization, image segmentation has to be applied, which requires clear boundaries to be present between different features [55]. The PCE-CT imaging mode appears, therefore, to be perfectly suited for the purpose, as the interfaces between RADs and TDs are accentuated and easy to identify, especially when compared to absorption images (Figure 4).

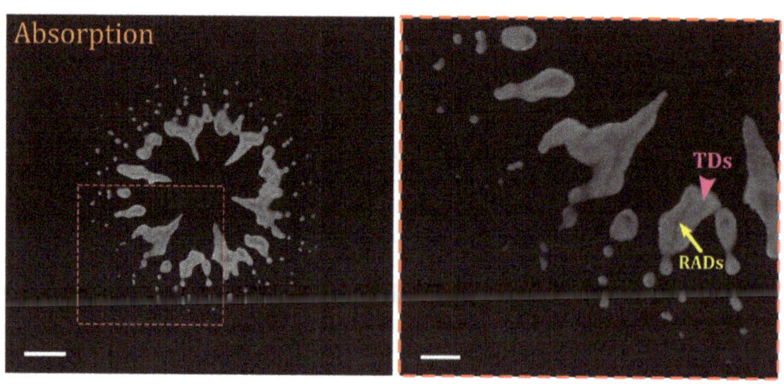

Figure 4. *Cont.*

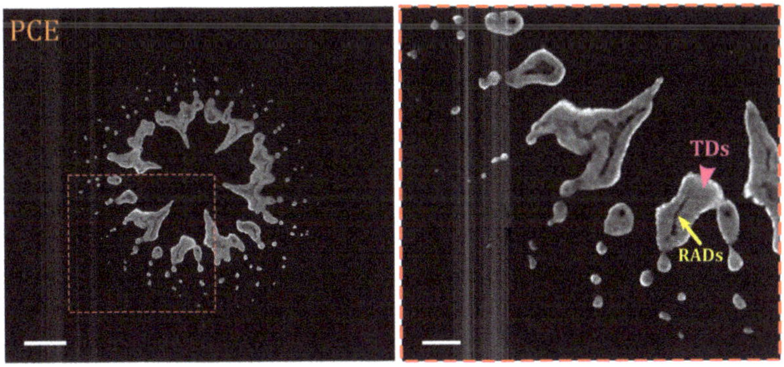

Figure 4. Typical coral skeleton details observed in cross-sectional tomography slices and comparison between absorption and PCE imaging modes. Differences in X-ray absorption within the skeleton reveal variations in both mineral distribution and density between RADs (darker regions correspond to lower density) and TDs (lighter regions, higher density); the red marked insets are shown on the right. Strongly defined boundaries can be observed between RADs and TDs in slices from PCE-CT datasets, whereas absorption images provide a blurry, poorly distinguishable separation between the two skeletal regions. Scale bars: 300 μm (**left** images), 100 μm (**right** images).

3.2. Segmentation of RADs and TDs

Conventional segmentation methods typically rely on choosing cut-off gray-values and are often based on the voxel intensity histograms. By applying a thresholding, the image can be segmented into the features that are below or above defined threshold values [56]. A gray-value histogram of an example coral skeleton shows that no clear distinction exists in gray-values corresponding TDs and RADs, as there is an overlap between the intensity ranges (for example, see gray-values of the margin of TDs facing the exterior of the skeleton and facing the RADs in Figure 5a,c). Phase-contrast images in fact are edge enhanced, meaning that both the brightest and lowest intensities of the image are found in the margins delineating the edges and internal boundaries of different densities and interfaces in the sample, and meaning that different features of interest may have the same "color".

Although PCE-CT imaging creates a sharp visual discrimination of RADs and TDs, a simple "binary" threshold applied to this type of image does not yield precise segmentation and reliable distinction between the two skeletal regions, due to the overlap between the intensity range of RADs and the TDs margins. This is shown by the mismatches between the distribution of RADs and TDs as observed in the reconstructed image and the distribution outlined by simple thresholding (Figure 5a,b). Thus, segmentation and ultimately quantification and 3D visualization of RADs and TDs require the use of a more sophisticated technique. Figure 6a depicts results obtained by the automatic segmentation performed using DL networks, which reveals a far better distinction of RADs and TDs within the recruit skeletons. The impressive separation between features becomes evident within the complex shapes of RADs across the skeleton (Video S1), and provides a means for overcoming the limitations imposed by the overlap in gray-values intensity ranges. Such detail-sensitive segmentation of RADs and TDs across an entire stack of slices ultimately allows for accurate volumetric quantification and 3D reproduction of the entire recruit's skeleton, including fine details of local thickness within RADs (Figure 6b, Video S1).

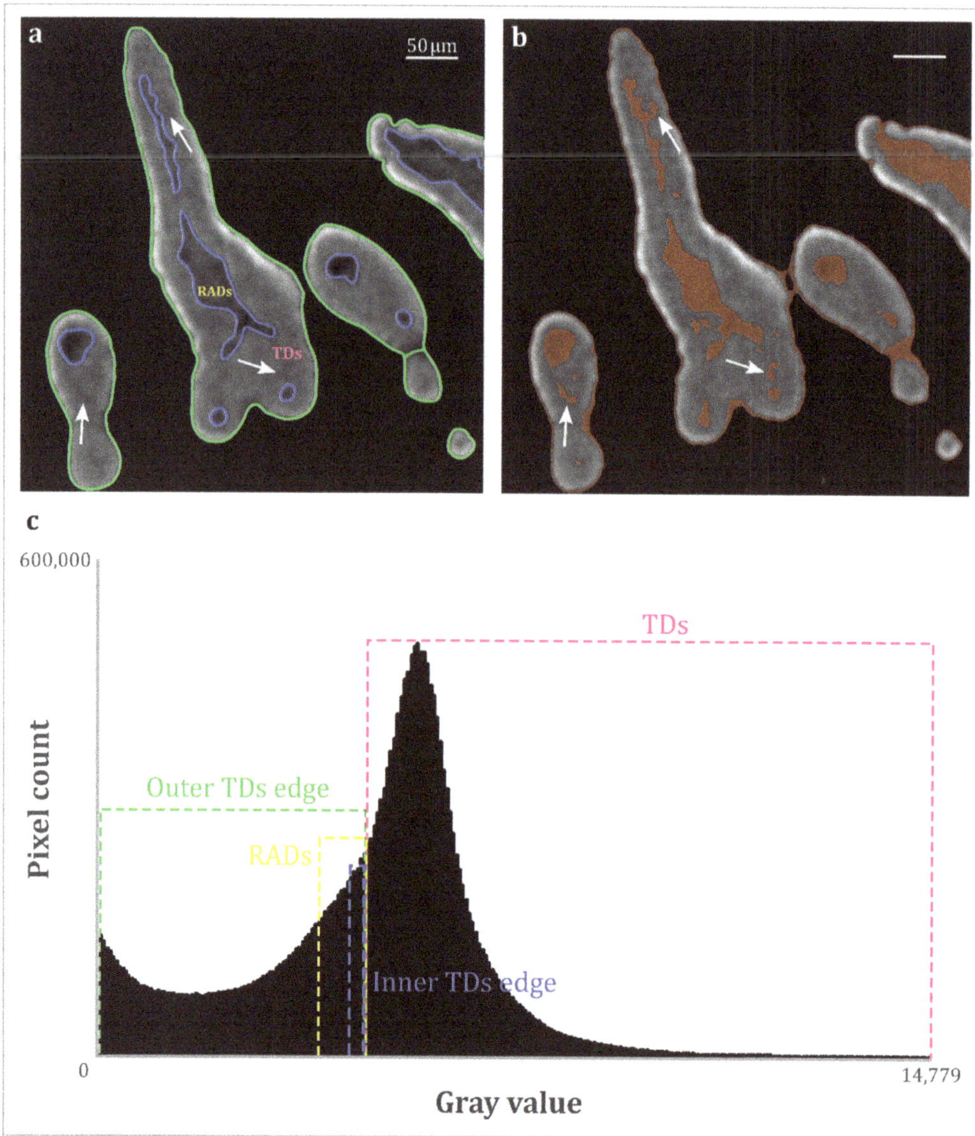

Figure 5. Slices and typical histogram of gray-values in PCE image and quality of non-DL-based image segmentation. (**a**) Close-up of the skeleton, showing RADs (dark gray areas), TDs (light gray areas), and delineating the outer and inner edges of TDs (green and pink lines, respectively). (**b**) Result of a simple threshold applied to segment RADs (RADs segmented by the thresholding method are colored in red). Note that, compared to their actual shape in the ground truth image (**a**), RADs are not precisely segmented (see arrows for example mismatch, where TDs regions are classified as RADs) and they are not distinguished from the outer edge of TDs, which are also thresholded as RADs by this method. (**c**) Gray-value histogram for an example tomographic stack of an imaged coral skeleton. The gray-value ranges for all regions depicted in (**a**) are highlighted in different colors. Note the overlap between the gray-value ranges of RADs and of the inner/outer edges of TDs, which makes it impossible to reliably segment these regions by simple threshold, as demonstrated in (**b**).

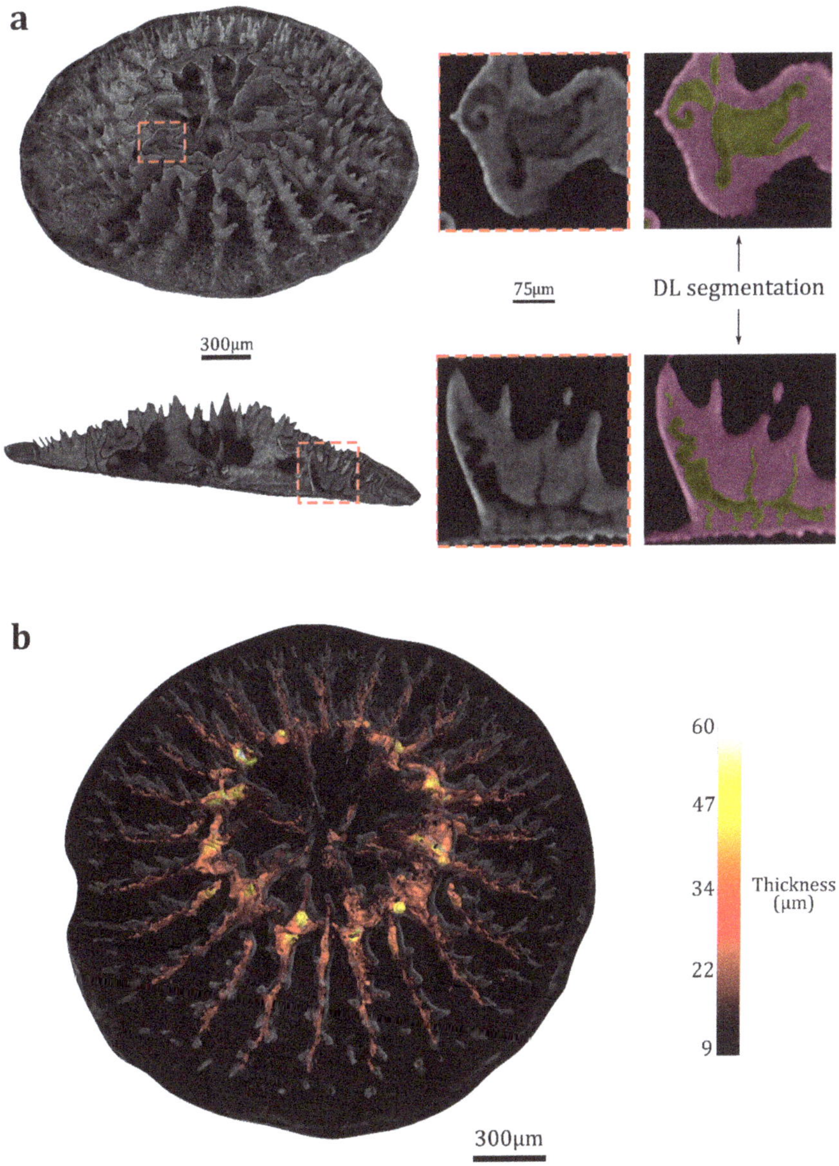

Figure 6. Deep-Learning-based image segmentation and quantitative 3D reproduction of segmented RADs and TDs. (**a**) Two example 3D renderings of a primary polyp skeleton virtually sliced in the transverse plane (upper row) and in the longitudinal plane (bottom row). Red dotted lines show enlargements of the sliced skeleton that have been segmented using DL (segmented TDs and RADs are colored in pink and yellow, respectively). DL, Deep Learning. (**b**) Segmentation of RADs and TDs across the entire stack of slices allows 3D separation of features and quantification, including details of local thickness (µm).

3.3. Evaluation of the DL-Based Segmentation Performance

Figures 6 and 7c,d show the intricate details made available due to the DL segmentation based on training an independent network (e.g., network A trained and applied on dataset A). Visual inspections of the segmentation outcomes reveal that applying a network optimized on a single tomographic dataset does not always provide a precise segmentation of other datasets (cross segmentation), as demonstrated by small mismatches between the real arrangement of RADs and TDs (as observed in "ground truth" images, Figure 7a,b) and the distribution outlined by the DL segmentation (Figure 7e,f).

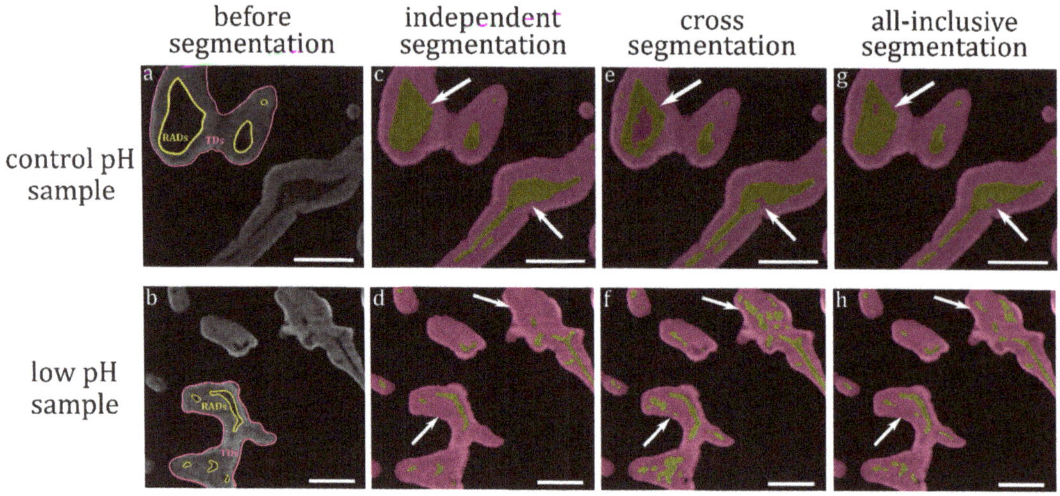

Figure 7. Evaluation of the DL-based segmentation performance. (**a**,**b**) Details of the skeletal structures in representative coral recruits from the control and low pH conditions before applying the DL-based segmentation (ground truth images). RADs and TDs in example areas have been manually delineated in yellow and pink, respectively. (**c**,**d**) Segmentation of RADs and TDs based on training a network independently per each single dataset. Specifically, (**c**) shows the outcome of the segmentation of network A trained and applied on dataset A (a representative control pH sample), whereas (**d**) shows the outcome of the segmentation of network X trained and applied on dataset X (a representative low pH sample). (**e**) Segmentation obtained by employing a DL network that was trained using dataset B (another representative control pH sample) and applied to dataset A. The white arrows point to mismatches in the segmentation as compared to the more accurate segmentation in (**c**). Note how RAD regions have been segmented as TDs. (**f**) Segmentation obtained by training the DL network on dataset A and using it to segment dataset X. The white arrows point to mismatches in the segmentation as compared to the correct segmentation in (**d**). (**g**,**h**) Segmentation obtained by employing a DL network that was trained using all tomographic datasets (all 6 datasets from both control and pH conditions). White arrows identify example points where different segmentation is seen in (**g**) as compared to (**e**), and the better segmentation in (**h**) as compared to (**f**). Scale bars: 100 µm.

Such visual segmentation mismatches for the cross segmentation cases lead to significant differences in the mean per-slice percentage of RADs area over the total area of the skeleton, for both control pH dataset (mean per-slice difference of −1.60% ± 0.07 between the % RADs area/total area obtained with independent segmentation and the one obtained with cross segmentation; Wilcoxon test, W = −20060, $p < 0.0001$) (red dots compared to green dots in Figure 8a) and for low pH dataset (mean per-slice difference of +7.90% ± 0.38 between the % RADs area/total area obtained with independent segmentation and the one obtained with cross segmentation; Wilcoxon test, W = 13861, $p < 0.0001$) (purple dots compared to blue dots in Figure 8b). Importantly, however, for the normal pH, the differences are less than 2%, which is impressive given the huge number of slices processed. The generation of such discrepancies may be due to the sensitivity of AI in segmenting images where there are large variations in the shape of the objects of interest between different datasets [51–53]. We indeed found substantial variations in the shape of skeletal features among datasets, as indicated by the different skewness values of the distributions of the RADs-TDs shape configurations (Figure 8c,d and Supplemental Figure S3). Curiously though, we did not find a significant correlation between these skewness values and the degree of overestimation or underestimation of the RADs/total skeleton area. This suggests that, on average, DL segmentation does not skew the data and is certainly no less accurate than other segmentation methods.

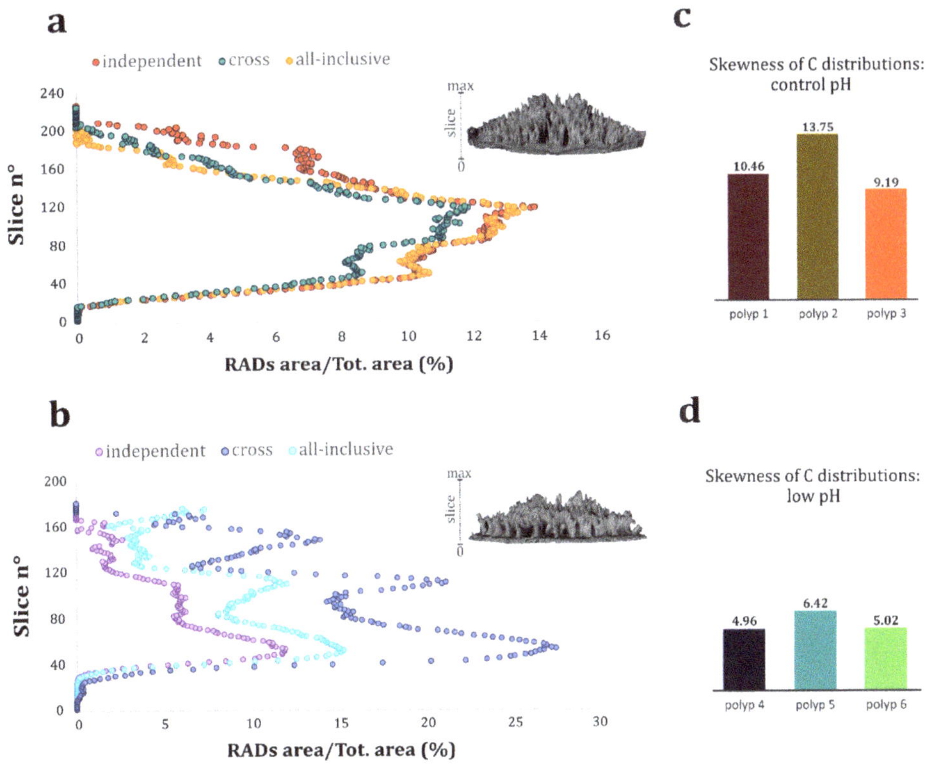

Figure 8. Evaluation of the DL-based segmentation performance based on the per-slice estimation of the area of the coral skeletal features and skewness of the distributions of the RADs-TDs shape configuration across samples. (**a**,**b**) Measured per slice area of RADs over the total area of the skeleton (RADs + TDs) estimated using independent, cross, and all-inclusive segmentations (as shown in Figure 7) in the case of a control pH sample (**a**) and a low pH sample (**b**). The slice number indicates the progression through the recruit's skeleton going from the base to the top, as illustrated by the coral skeletons on the right side of the graphs. Note how, for both control and low pH datasets, the per-slice area of the RADs obtained with all-inclusive segmentation gets closer to the value obtained with independent segmentation (red dots versus yellow dots and purple dots versus light blue dots, respectively) compared to the per-slice area obtained with cross segmentation (red dots versus green dots and purple dots versus blue dots, respectively). (**c**,**d**) Skewness of the distributions of the per-slice RADs-TDs shape configuration in control pH (**c**) and low pH (**d**) samples (see also Supplemental Figure S3). The ratio of the RADs perimeter/area over the TDs perimeter/area ("C") was computed as an estimate of the relative shape of the skeletal features across the recruit skeletons. The skewness values are used as a metric to evaluate the degree of the RADs-TDs shape variation across datasets.

Notably, differences in the skewness of distributions are particularly evident when comparing datasets across pH conditions. Low pH datasets in fact have a significantly lower mean skewness (5.47 ± 0.48) compared to control pH dataset (mean of 11.13 ± 1.36; Unpaired t test, t = 3.922, p = 0.017) (Figure 8c,d), suggesting an underlying shaping effect of acidic seawater on the RADs-TDs distribution across the recruits skeleton, requiring further in-depth investigation.

Further visual evaluation of the different segmentation outcomes reveals that applying a network trained using all tomographic datasets (named all-inclusive segmentation) provides a much improved segmentation performance compared to the cross segmentation (Figure 7e–h). These observations are supported by the measurements of the mean per-slice percentage of RADs area over the total area of the skeleton (Figure 8a,b). For both control and low pH datasets, the all-inclusive segmentation yielded a significantly better estimation of such percent area than the one produced by the cross segmentation (for the control pH dataset: mean per-slice difference of −0.95% ± 0.10 between the % RADs area/total area obtained with independent segmentation and the one obtained with all-inclusive segmentation, Wilcoxon test, W = −13081, p < 0.0001; for the low pH dataset: mean per-slice difference of +2.65% ± 0.15 between the % RADs area/total area obtained with independent segmentation and the one obtained with all-inclusive segmentation, Wilcoxon test, W = −13651, p < 0.0001)(see red dots compared to yellow dots in Figure 8a and purple dots compared to light blue dots in Figure 8b). These results indicate that variations in the coral geometries themselves increase the variability of shapes encountered by the model during training, which in turn substantially increases the robustness of the DL-based segmentation performance.

4. Discussion

Combining PCE-CT and AI-based image segmentation, we have developed a highly sensitive methodology to obtain quantitative volumetric information about RADs and TDs, the two main skeleton domains of reef-building corals (Figure 6, Video S1). As is the case for many tomographic analyses, image segmentation of the coral is a necessary step for quantitative analytical investigations of the datasets. Our work provides a pipeline that allows to extract the regions corresponding to volumes and densities of interest and to obtain their detailed 3D geometry quantification. Image contrast differences strongly influence the segmentation performance, as they determine how well objects of interests are differentiated by both human and machine vision, thus improving the object perceptual quality [57,58]. In the present study, we show that PCE-CT yields the needed high contrast difference that is essential to visually identify the fine intricate interwoven structures of TDs and RADs within the intact coral skeleton. Indeed, PCE has proven to be a powerful tomographic method in both biomedical research and geological sciences to increase image contrast and enhance image segmentation performance [59].

Although PCE-CT imaging greatly facilitates the visual distinction between RADs and TDs by accentuating their µm-sized boundaries, it introduces difficulties to precisely segment the two skeletal regions using simple segmentation techniques. Specifically, edge enhancement and strong artefacts totally preclude any phase retrieval such as the paganin method [60]. For many datasets, thresholding is the simplest and more common method of segmenting images, which assumes that it is possible to find a single threshold value coinciding with the grayscale intensity of the pixels that correspond to the same structure in the volumetric image [56]. However, accurate image thresholding is only possible in cases where the different features in an image have clearly different peaks in the gray-value histogram. The significant overlap observed between the gray-value ranges of RADs and of the TDs edges makes it impossible to achieve a complete segmentation of the skeletal structures using a simple threshold (Figure 5). It is for this reason that DL algorithms are of great interest, as they may help overcome this difficulty by iteratively learning from data, allowing computers to digitally dissect images through automatic detection and classification of pixels belonging to the structures of interest [61]. In spite of the high complexity of the RADs structure across the recruit skeletons, the automatic segmentation performed by training DL networks delineated and distinguished RADs and TDs with great detail, an essential step on the path for volumetric quantification and visualization (Figure 6 and Video S1). Though computationally intensive, automated image segmentation has been widely reported in recent years for biomedical image segmentation (reviewed in [62]) as well as for mineral characterization [30–33]. However, up to now, the use of AI for retrieval of biologically interpretable results in the context of coral skeleton morphology has never been reported.

The impressive, robust performance of the DL-based segmentation in identifying and distinguish RADs and TDs across the coral skeleton particularly applies in the case of DL networks optimized for single datasets independently. In contrast, a network that was trained on one specific tomographic dataset yields severe mismatches when applied to segmentation of a different dataset, presumably because it is presented with previously unencountered shape configurations of the target skeletal features (Figures 7 and 8, Supplemental Figure S3). This suggests that the variability in coral geometries must be considered and possibly quantified, when generalizing findings, e.g., TDs or RADs morphologies. Furthermore, this suggests that the performance of AI in segmenting coral skeletal structures is quite sensitive to the large variability of the shape of RADs and TDs that exist among different recruits. Such shape variability is especially evident when comparing control pH recruits to low pH recruits, suggesting a morphologically-altering effect of OA on coral skeletal features that will require further in-depth examination to be deciphered. We speculate that the significant morphological differences tip the ratio between TDs and RADs in the skeletons in such a way that AI needs much more information to fully characterize the effects of OA. Note that the number of regions and slices presented to the DL for training in the normal pH and OA affected recruits was identical. This clearly indicates that normal pH corals are more similar to one another than to the pH affected corals.

Indeed, varying appearance of target features poses a great challenge in image segmentation (reviewed in [54]). For example in medical research, there can be wide differences between different medical images that affect the adaptability of DL network models during segmentation [62,63]. This is mainly due to the heterogeneous appearance of target organs for segmentation, which may vary hugely in size, shape and location from one patient to the other [64]. Another aspect that may restrict the performance of DL-based image segmentation is the limited diversity of data available to train a network [54,62,63]. Here, we show that widening the spectrum of possible appearances of the target skeletal features to be recognized using AI across multiple datasets considerably improves the DL-based segmentation performance (Figures 7 and 8). This showcases the power of AI when fed with diverse training data, and the potential of application across multiple tomographic datasets.

A commonly adopted method to generate new samples and increase the size of the training datasets is data augmentation, which is the application of a set of image transformations (e.g., rotation or flipping) to the dataset [65]. In this way, previously unencountered target features can potentially be approximated by the network. However, this method also adds sources of error and variability to the general methodological framework, as so far there is no strict consensus on which transformation to introduce and generally the transformation parameters are inconstantly chosen [54]. In this regard, we aim at improving the performance and generalizability of DL models across diverse coral datasets by collecting a larger diverse array of coral tomography data. If such datasets are made publicly available through, for example, open-access online databases, they will form a source of data that can be used to train robust, efficient, and generalizable DL models for use in coral research. It should be noted that image processing and segmentation of entire coral polyps with resolutions of ~2 μm and, in particular, using DL networks require very large computation memory and power, for which high-performance workstations may not be ubiquitously available. Yet, providing tools to help researchers analyze and utilize such data is essential, thus forming a data-collection-analysis feedback loop. Our work presented here is a first step on this ambitious path.

5. Conclusions

To the best of our knowledge, this is the first study that applied AI approaches to investigate the internal 3D geometry of RADs and TDs in coral skeletons based on high-sensitivity, non-destructive PCE-CT data. The approach presented here opens the possibility of using AI to reconstruct and quantitatively analyze the internal skeletal network of reef building corals. In particular, such technology has the potential to improve ecological monitoring programs, by assessing coral skeletal anatomy and providing morphological descriptors in order to taxonomize hard corals to genus and species levels. In addition, it can expand our knowledge on coral biomineralization by providing a means of characterizing the microscale interfaces within the different coral skeletal regions. Lastly, such a tool allows us to study the dynamics of coral skeletal formation under a variety of environmental conditions, including predicted OA scenarios.

In the last decades we witnessed a significant decline in coral reefs all over the world, and as global climate changes are becoming more severe, the need to protect coral reefs is becoming urgent. Hence, applying materials science contrast-enhanced methods to structural research in corals would be of great benefit for the development of highly targeted restoration strategies, as it would critically expand our knowledge on the effects of environmental change on coral skeleton development.

Supplementary Materials: The following are available online at https://www.mdpi.com/article/10.3390/jmse10030391/s1, Figure S1: Coral sample fixed in the sample holder in front of the X-ray source. Figure S2: Density variation among RADs and TDs. Figure S3: Skewness of the distributions of the per-slice RADs-TDs shape configuration across tomographic datasets. Table S1: Parameters of the carbonate chemistry across experimental pH conditions. Video S1: 3D view of coral skeletal structures before and after DL-based-segmentation.

Author Contributions: Conceptualization, F.S., T.M. and P.Z.; Formal analysis, F.S.; Funding acquisition, T.M. and P.Z.; Investigation, F.S., K.S. and P.Z.; Methodology, F.S. and P.Z.; Project administration, T.M. and P.Z.; Supervision, T.M. and P.Z.; Visualization, F.S.; Writing—original draft, F.S. and P.Z.; Writing—review & editing, F.S., K.S., T.M. and P.Z. All authors have read and agreed to the published version of the manuscript.

Funding: This project has received funding from GIF, the German-Israeli Foundation for Scientific Research and Development (I-1496-302.8), from the Israeli Binational Science Foundation (BSF 2016321 to T.M.) and from the European Research Council (ERC) under the European Union's Horizon 2020 research and innovation programme (grant agreement no. 755876 to T.M.). The experiment was performed in a controlled aquarium system which was funded by Institutional ISF grants 2288/16.

Institutional Review Board Statement: Not applicable.

Informed Consent Statement: Not applicable.

Data Availability Statement: All data needed to evaluate the conclusions in the paper are present in the paper and/or the Supplemental Materials. Raw X-ray μCT data are available upon request from the corresponding authors.

Acknowledgments: We thank the HZB for granting us the use of the BAMline, specifically Henning Markötter and Michael Sintschuk for beamtime setup. We also thank Maayan Neder for her help in collecting coral larvae from the wild, Tal Zoquin for his assistance with setting up the OA experiment and Ole Lenz for his help in establishing the reconstruction pipeline and guidance in dataset processing.

Conflicts of Interest: The authors declare no conflict of interest.

References

1. Brahmi, C.; Domart-Coulon, I.; Rougée, L.; Pyle, D.G.; Stolarski, J.; Mahoney, J.J.; Richmond, R.H.; Ostrander, G.K.; Meibom, A. Pulsed 86Sr-labeling and NanoSIMS imaging to study coral biomineralization at ultra-structural length scales. *Coral Reefs* **2012**, *31*, 741–752. [CrossRef]
2. Cuif, J.-P.; Dauphin, Y. The two-step mode of growth in the scleractinian coral skeletons from the micrometre to the overall scale. *J. Struct. Biol.* **2005**, *150*, 319–331. [CrossRef] [PubMed]

3. Stolarski, J. Three-dimensional micro-and nanostructural characteristics of the scleractinian coral skeleton: A biocalcification proxy. *Acta Palaeontol. Pol.* **2003**, *48*, 4.
4. Meibom, A.; Cuif, J.-P.; Hillion, F.; Constantz, B.R.; Juillet-Leclerc, A.; Dauphin, Y.; Watanabe, T.; Dunbar, R.B. Distribution of magnesium in coral skeleton: MG microdistribution in coral. *Geophys. Res. Lett.* **2004**, *31*. [CrossRef]
5. Von Euw, S.; Zhang, Q.; Manichev, V.; Murali, N.; Gross, J.; Feldman, L.C.; Gustafsson, T.; Flach, C.; Mendelsohn, R.; Falkowski, P.G. Biological control of aragonite formation in stony corals. *Science* **2017**, *356*, 933–938. [CrossRef]
6. Mass, T.; Giuffre, A.J.; Sun, C.-Y.; Stifler, C.A.; Frazier, M.J.; Neder, M.; Tamura, N.; Stan, C.V.; Marcus, M.A.; Gilbert, P.U.P.A. Amorphous calcium carbonate particles form coral skeletons. *Proc. Natl. Acad. Sci. USA* **2017**, *114*, E7670–E7678. [CrossRef] [PubMed]
7. Neder, M.; Laissue, P.P.; Akiva, A.; Akkaynak, D.; Albéric, M.; Spaeker, O.; Politi, Y.; Pinkas, I.; Mass, T. Mineral formation in the primary polyps of pocilloporoid corals. *Acta Biomater.* **2019**, *96*, 631–645. [CrossRef]
8. Cuif, J.-P.; Dauphin, Y. Microstructural and physico-chemical characterization of 'centers of calcification' in septa of some Recent scleractinian corals. *Paläontol. Z.* **1998**, *72*, 257–269. [CrossRef]
9. Frankowiak, K.; Mazur, M.; Gothmann, A.M.; Stolarski, J. Diagenetic alteration of triassic coral from the aragonite konservat-lagerstatte in alakir cay, turkey: Implications for geochemical measurements. *Palaios* **2013**, *28*, 333–342. [CrossRef]
10. Mass, T.; Drake, J.L.; Peters, E.C.; Jiang, W.; Falkowski, P.G. Immunolocalization of skeletal matrix proteins in tissue and mineral of the coral Stylophora pistillata. *Proc. Natl. Acad. Sci. USA* **2014**, *111*, 12728–12733. [CrossRef]
11. Sugiura, M.; Yasumoto, K.; Iijima, M.; Oaki, Y.; Imai, H. Morphological study of fibrous aragonite in the skeletal framework of a stony coral. *CrystEng. Comm.* **2021**, *23*, 3693–3700. [CrossRef]
12. Benzerara, K.; Menguy, N.; Obst, M.; Stolarski, J.; Mazur, M.; Tylisczak, T.; Brown, G.E.; Meibom, A. Study of the crystallographic architecture of corals at the nanoscale by scanning transmission X-ray microscopy and transmission electron microscopy. *Ultramicroscopy* **2011**, *111*, 1268–1275. [CrossRef] [PubMed]
13. Brahmi, C.; Meibom, A.; Smith, D.C.; Stolarski, J.; Auzoux-Bordenave, S.; Nouet, J.; Doumenc, D.; Djediat, C.; Domart-Coulon, I. Skeletal growth, ultrastructure and composition of the azooxanthellate scleractinian coral Balanophyllia regia. *Coral Reefs* **2010**, *29*, 175–189. [CrossRef]
14. Cohen, A.; Holcomb, M. Why Corals Care About Ocean Acidification: Uncovering the Mechanism. *Oceanograpgy* **2009**, *22*, 118–127. [CrossRef]
15. Erez, J.; Reynaud, S.; Silverman, J.; Schneider, K.; Allemand, D. Coral Calcification Under Ocean Acidification and Global Change. In *Coral Reefs: An Ecosystem in Transition*; Dubinsky, Z., Stambler, N., Eds.; Springer Netherlands: Dordrecht, The Netherlands, 2011; pp. 151–176. ISBN 978-94-007-0113-7.
16. Kleypas, J.A.; Langdon, C. Coral reefs and changing seawater carbonate chemistry. In *Coastal and Estuarine Studies*; Phinney, J.T., Hoegh-Guldberg, O., Kleypas, J., Skirving, W., Strong, A., Eds.; American Geophysical Union: Washington, DC, USA, 2006; Volume 61, pp. 73–110. ISBN 978-0-87590-359-0.
17. Marubini, F.; Ferrier–Pages, C.; Cuif, J. Suppression of skeletal growth in scleractinian corals by decreasing ambient carbonate-ion concentration: A cross-family comparison. *Proc. R. Soc. Lond. B* **2003**, *270*, 179–184. [CrossRef]
18. Mollica, N.R.; Guo, W.; Cohen, A.L.; Huang, K.-F.; Foster, G.L.; Donald, H.K.; Solow, A.R. Ocean acidification affects coral growth by reducing skeletal density. *Proc. Natl. Acad. Sci. USA* **2018**, *115*, 1754–1759. [CrossRef]
19. Scucchia, F.; Malik, A.; Zaslansky, P.; Putnam, H.M.; Mass, T. Combined responses of primary coral polyps and their algal endosymbionts to decreasing seawater pH. *Proc. R. Soc. B* **2021**, *288*, 20210328. [CrossRef]
20. Cohen, A.L.; McCorkle, D.C.; de Putron, S.; Gaetani, G.A.; Rose, K.A. Morphological and compositional changes in the skeletons of new coral recruits reared in acidified seawater: Insights into the biomineralization response to ocean acidification. *Geochem. Geophys. Geosyst.* **2009**, *10*. [CrossRef]
21. Nothdurft, L.D.; Webb, G.E. Microstructure of Common Reef-Building Coral Genera *Acropora*, *Pocillopora*, *Goniastrea* and *Porites*: Constraints on Spatial Resolution in Geochemical Sampling. *Facies* **2007**, *53*, 1–26. [CrossRef]
22. Cnudde, V.; Boone, M.N. High-resolution X-ray computed tomography in geosciences: A review of the current technology and applications. *Earth-Sci. Rev.* **2013**, *123*, 1–17. [CrossRef]
23. Kyle, J.R.; Ketcham, R.A. Application of high resolution X-ray computed tomography to mineral deposit origin, evaluation, and processing. *Ore Geol. Rev.* **2015**, *65*, 821–839. [CrossRef]
24. Cloetens, P.; Pateyron-Salomé, M.; Buffière, J.Y.; Peix, G.; Baruchel, J.; Peyrin, F.; Schlenker, M. Observation of microstructure and damage in materials by phase sensitive radiography and tomography. *J. Appl. Phys.* **1997**, *81*, 5878–5886. [CrossRef]
25. Baruchel, J. *X-ray Tomography in Material Science*; Hermes Science: Paris, France, 2000; ISBN 978-2-7462-0115-6.
26. De Albuquerque, V.H.C.; Cortez, P.C.; de Alexandria, A.R.; Tavares, J.M.R.S. A new solution for automatic microstructures analysis from images based on a backpropagation artificial neural network. *Nondestruct. Test. Eval.* **2008**, *23*, 273–283. [CrossRef]
27. De Albuquerque, V.H.C.; Tavares, J.M.R.S.; Durão, L.M.P. Evaluation of Delamination Damage on Composite Plates using an Artificial Neural Network for the Radiographic Image Analysis. *J. Compos. Mater.* **2010**, *44*, 1139–1159. [CrossRef]
28. Iglesias, J.C.Á.; Santos, R.B.M.; Paciornik, S. Deep learning discrimination of quartz and resin in optical microscopy images of minerals. *Miner. Eng.* **2019**, *138*, 79–85. [CrossRef]
29. Buyya, R.; Calheiros, R.N.; Dastjerdi, A.V. (Eds.) *Big Data: Principles and Paradigms*; Elsevier/Morgan Kaufmann: Cambridge, MA, USA, 2016; ISBN 978-0-12-805394-2.

30. Chauhan, S.; Rühaak, W.; Anbergen, H.; Kabdenov, A.; Freise, M.; Wille, T.; Sass, I. Phase segmentation of X-ray computer tomography rock images usingmachine learning techniques: An accuracy and performancestudy. *Solid Earth* **2016**, *7*, 1125–1139. [CrossRef]
31. Chauhan, S.; Rühaak, W.; Khan, F.; Enzmann, F.; Mielke, P.; Kersten, M.; Sass, I. Processing of rock core microtomography images: Using seven different machine learning algorithms. *Comput. Geosci.* **2016**, *86*, 120–128. [CrossRef]
32. Varfolomeev, I.; Yakimchuk, I.; Safonov, I. An Application of Deep Neural Networks for Segmentation of Microtomographic Images of Rock Samples. *Computers* **2019**, *8*, 72. [CrossRef]
33. Wang, Y.; Lin, C.L.; Miller, J.D. Improved 3D image segmentation for X-ray tomographic analysis of packed particle beds. *Miner. Eng.* **2015**, *83*, 185–191. [CrossRef]
34. Van Vuuren, D.P.; Edmonds, J.; Kainuma, M.; Riahi, K.; Thomson, A.; Hibbard, K.; Hurtt, G.C.; Kram, T.; Krey, V.; Lamarque, J.-F.; et al. The representative concentration pathways: An overview. *Clim. Chang.* **2011**, *109*, 5–31. [CrossRef]
35. Pörtner, H.-O.; Roberts, D.C.; Masson-Delmotte, V.; Zhai, P.; Tignor, M.; Poloczanska, E.; Mintenbeck, K.; Alegría, A.; Nicolai, M.; Okem, A.; et al. (Eds.) *Special Report on the Ocean and Cryosphere in a Changing Climate*; IPCC: Geneva, Switzerland, 2019.
36. Goodbody-Gringley, G. Diel Planulation by the Brooding Coral Favia Fragum (Esper, 1797). *J. Exp. Mar. Biol. Ecol.* **2010**, *389*, 70–74. [CrossRef]
37. Goodbody-Gringley, G.; Wong, K.H.; Becker, D.M.; Glennon, K.; de Putron, S.J. Reproductive Ecology and Early Life History Traits of the Brooding Coral, *Porites astreoides*, from Shallow to Mesophotic Zones. *Coral Reefs* **2018**, *37*, 483–494. [CrossRef]
38. Pierrot, D.; Lewis, E.; Wallace, D.W.R. *MS Excel Program Developed for CO2 System Calculations*; ORNL/CDIAC-105a; Carbon Dioxide Information Analysis Center, Oak Ridge National Laboratory, US Department of Energy: Oak Ridge, TN, USA, 2006; Volume 10.
39. Mehrbach, C.; Culberson, C.H.; Hawley, J.E.; Pytkowicx, R.M. Measurement of the apparent dissociation constants of carbonic acid in seawater at atmospheric pressure 1. *Limnol. Oceanogr.* **1973**, *18*, 897–907. [CrossRef]
40. Dickson, A.G.; Millero, F.J. A comparison of the equilibrium constants for the dissociation of carbonic acid in seawater media. *Deep. Sea Res. Part A Oceanogr. Res. Pap.* **1987**, *34*, 1733–1743. [CrossRef]
41. Görner, W.; Hentschel, M.P.; Müller, B.R.; Riesemeier, H.; Krumrey, M.; Ulm, G.; Diete, W.; Klein, U.; Frahm, R. BAMline: The first hard X-ray beamline at BESSY II. *Nucl. Instrum. Methods Phys. Res. Sect. A* **2001**, *467–468*, 703–706. [CrossRef]
42. Rack, A.; Zabler, S.; Müller, B.R.; Riesemeier, H.; Weidemann, G.; Lange, A.; Goebbels, J.; Hentschel, M.; Görner, W. High resolution synchrotron-based radiography and tomography using hard X-rays at the BAMline (BESSY II). *Nucl. Instrum. Methods Phys. Res. Sect. A* **2008**, *586*, 327–344. [CrossRef]
43. Zaslansky, P.; Fratzl, P.; Rack, A.; Wu, M.-K.; Wesselink, P.R.; Shemesh, H. Identification of root filling interfaces by microscopy and tomography methods: Microtomography and microscopy observations of root fillings. *Int. Endod. J.* **2011**, *44*, 395–401. [CrossRef]
44. Reznikov, N.; Buss, D.J.; Provencher, B.; McKee, M.D.; Piché, N. Deep learning for 3D imaging and image analysis in biomineralization research. *J. Struct. Biol.* **2020**, *212*, 107598. [CrossRef]
45. Schindelin, J.; Arganda-Carreras, I.; Frise, E.; Kaynig, V.; Longair, M.; Pietzsch, T.; Preibisch, S.; Rueden, C.; Saalfeld, S.; Schmid, B.; et al. Fiji: An open-source platform for biological-image analysis. *Nat. Methods* **2012**, *9*, 676–682. [CrossRef]
46. Makovetsky, R.; Piche, N.; Marsh, M. Dragonfly as a Platform for Easy Image-based Deep Learning Applications. *Microsc. Microanal.* **2018**, *24*, 532–533. [CrossRef]
47. Tan, Y. Applications. In *Gpu-Based Parallel Implementation of Swarm Intelligence Algorithms*; Elsevier: Amsterdam, The Netherlands, 2016; pp. 167–177. ISBN 978-0-12-809362-7.
48. Ronneberger, O.; Fischer, P.; Brox, T. U-Net: Convolutional Networks for Biomedical Image Segmentation. In *Medical Image Computing and Computer-Assisted Intervention—MICCAI 2015*; Navab, N., Hornegger, J., Wells, W.M., Frangi, A.F., Eds.; Lecture Notes in Computer Science; Springer International Publishing: Cham, Switzerland, 2015; Volume 9351, pp. 234–241. ISBN 978-3-319-24573-7.
49. Hamwood, J.; Alonso-Caneiro, D.; Read, S.A.; Vincent, S.J.; Collins, M.J. Effect of patch size and network architecture on a convolutional neural network approach for automatic segmentation of OCT retinal layers. *Biomed. Opt. Express* **2018**, *9*, 3049. [CrossRef]
50. Sudre, C.H.; Li, W.; Vercauteren, T.; Ourselin, S.; Jorge Cardoso, M. Generalised Dice Overlap as a Deep Learning Loss Function for Highly Unbalanced Segmentations. In *Deep Learning in Medical Image Analysis and Multimodal Learning for Clinical Decision Support*; Cardoso, M.J., Arbel, T., Carneiro, G., Syeda-Mahmood, T., Tavares, J.M.R.S., Moradi, M., Bradley, A., Greenspan, H., Papa, J.P., Madabhushi, A., et al., Eds.; Lecture Notes in Computer Science; Springer International Publishing: Cham, Switzerland, 2017; Volume 10553, pp. 240–248. ISBN 978-3-319-67557-2.
51. Khalifa, F.; El-Baz, A.; Gimel'farb, G.; Ouseph, R.; El-Ghar, M.A. Shape-Appearance Guided Level-Set Deformable Model for Image Segmentation. In Proceedings of the 2010 20th International Conference on Pattern Recognition, Istanbul, Turkey, 23–26 August 2010; pp. 4581–4584.
52. El-Baz, A.; Gimel'farb, G. Robust image segmentation using learned priors. In Proceedings of the 2009 IEEE 12th International Conference on Computer Vision, Kyoto, Japan, 29 September–2 October 2009; pp. 857–864.
53. Chen, F.; Yu, H.; Hu, R.; Zeng, X. Deep Learning Shape Priors for Object Segmentation. In Proceedings of the 2013 IEEE Conference on Computer Vision and Pattern Recognition, Portland, OR, USA, 23–28 June 2013; pp. 1870–1877.

54. Renard, F.; Guedria, S.; Palma, N.D.; Vuillerme, N. Variability and reproducibility in deep learning for medical image segmentation. *Sci. Rep.* **2020**, *10*, 13724. [CrossRef] [PubMed]
55. Sharon, E.; Brandt, A.; Basri, R. Segmentation and boundary detection using multiscale intensity measurements. In Proceedings of the 2001 IEEE Computer Society Conference on Computer Vision and Pattern Recognition, Kauai, HI, USA, 8–14 December 2001; Volume 1.
56. Guntoro, P.I.; Ghorbani, Y.; Koch, P.-H.; Rosenkranz, J. X-ray Microcomputed Tomography (μCT) for Mineral Characterization: A Review of Data Analysis Methods. *Minerals* **2019**, *9*, 183. [CrossRef]
57. Campos, G.F.C.; Mastelini, S.M.; Aguiar, G.J.; Mantovani, R.G.; de Melo, L.F.; Barbon, S. Machine learning hyperparameter selection for Contrast Limited Adaptive Histogram Equalization. *J. Image Video Proc.* **2019**, *2019*, 59. [CrossRef]
58. Saleem, A.; Beghdadi, A.; Boashash, B. Image fusion-based contrast enhancement. *J. Image Video Proc.* **2012**, *2012*, 10. [CrossRef]
59. Withers, P.J.; Bouman, C.; Carmignato, S.; Cnudde, V.; Grimaldi, D.; Hagen, C.K.; Maire, E.; Manley, M.; Du Plessis, A.; Stock, S.R. X-ray computed tomography. *Nat. Rev. Methods Primers* **2021**, *1*, 18. [CrossRef]
60. Paganin, D.; Mayo, S.C.; Gureyev, T.E.; Miller, P.R.; Wilkins, S.W. Simultaneous phase and amplitude extraction from a single defocused image of a homogeneous object. *J. Microsc.* **2002**, *206*, 33–40. [CrossRef]
61. LeCun, Y.; Bengio, Y.; Hinton, G. Deep learning. *Nature* **2015**, *521*, 436–444. [CrossRef]
62. Liu, X.; Song, L.; Liu, S.; Zhang, Y. A Review of Deep-Learning-Based Medical Image Segmentation Methods. *Sustainability* **2021**, *13*, 1224. [CrossRef]
63. Hesamian, M.H.; Jia, W.; He, X.; Kennedy, P. Deep Learning Techniques for Medical Image Segmentation: Achievements and Challenges. *J. Digit. Imag.* **2019**, *32*, 582–596. [CrossRef]
64. Kamnitsas, K.; Ledig, C.; Newcombe, V.F.J.; Simpson, J.P.; Kane, A.D.; Menon, D.K.; Rueckert, D.; Glocker, B. Efficient multi-scale 3D CNN with fully connected CRF for accurate brain lesion segmentation. *Med. Image Anal.* **2017**, *36*, 61–78. [CrossRef] [PubMed]
65. Shorten, C.; Khoshgoftaar, T.M. A survey on Image Data Augmentation for Deep Learning. *J. Big Data* **2019**, *6*, 60. [CrossRef]

Journal of
Marine Science and Engineering

MDPI

Article

Ocean Warming Amplifies the Effects of Ocean Acidification on Skeletal Mineralogy and Microstructure in the Asterinid Starfish *Aquilonastra yairi*

Munawar Khalil [1,2,3,*], Steve S. Doo [1,4], Marleen Stuhr [1] and Hildegard Westphal [1,2,4]

1 Leibniz Centre for Tropical Marine Research (ZMT), 28359 Bremen, Germany; steve.doo@leibniz-zmt.de (S.S.D.); marleen.stuhr@leibniz-zmt.de (M.S.); hildegard.westphal@kaust.edu.sa (H.W.)
2 Faculty of Geosciences, University of Bremen, 28359 Bremen, Germany
3 Department of Marine Science, Faculty of Agriculture, Universitas Malikussaleh, Reuleut Main Campus, Aceh 24355, Indonesia
4 Physical Science and Engineering (PSE), Red Sea Research Center (RSRC), King Abdullah University of Science and Technology (KAUST), Thuwal 23955-6900, Saudi Arabia
* Correspondence: munawar.khalil@leibniz-zmt.de

Abstract: Ocean acidification and ocean warming compromise the capacity of calcifying marine organisms to generate and maintain their skeletons. While many marine calcifying organisms precipitate low-Mg calcite or aragonite, the skeleton of echinoderms consists of more soluble Mg-calcite. To assess the impact of exposure to elevated temperature and increased pCO_2 on the skeleton of echinoderms, in particular the mineralogy and microstructure, the starfish *Aquilonastra yairi* (Echinodermata: Asteroidea) was exposed for 90 days to simulated ocean warming (27 °C and 32 °C) and ocean acidification (455 µatm, 1052 µatm, 2066 µatm) conditions. The results indicate that temperature is the major factor controlling the skeletal Mg (Mg/Ca ratio and Mg_{norm} ratio), but not for skeletal Sr (Sr/Ca ratio and Sr_{norm} ratio) and skeletal Ca (Ca_{norm} ratio) in *A. yairi*. Nevertheless, inter-individual variability in skeletal Sr and Ca ratios increased with higher temperature. Elevated pCO_2 did not induce any statistically significant element alterations of the skeleton in all treatments over the incubation time, but increased pCO_2 concentrations might possess an indirect effect on skeletal mineral ratio alteration. The influence of increased pCO_2 was more relevant than that of increased temperature on skeletal microstructures. pCO_2 as a sole stressor caused alterations on stereom structure and degradation on the skeletal structure of *A. yairi*, whereas temperature did not; however, skeletons exposed to elevated pCO_2 and high temperature show a strongly altered skeleton structure compared to ambient temperature. These results indicate that ocean warming might exacerbate the skeletal maintaining mechanisms of the starfish in a high pCO_2 environment and could potentially modify the morphology and functions of the starfish skeleton.

Keywords: ocean acidification; ocean warming; echinoderm; starfish; mineralogy; skeleton; biomineralization

Citation: Khalil, M.; Doo, S.S.; Stuhr, M.; Westphal, H. Ocean Warming Amplifies the Effects of Ocean Acidification on Skeletal Mineralogy and Microstructure in the Asterinid Starfish *Aquilonastra yairi*. *J. Mar. Sci. Eng.* **2022**, *10*, 1065. https://doi.org/10.3390/jmse10081065

Academic Editor: Markes E. Johnson

Received: 31 May 2022
Accepted: 1 July 2022
Published: 3 August 2022

1. Introduction

The oceans are estimated to take up ~31% of the CO_2 increase that is currently observed [1]. This is known to lead to lowered pH values of the seawater, resulting in a reduced carbonate saturation state (Ω), and changes in carbonate–bicarbonate ion balance, recognized as ocean acidification (OA) [2,3]. These changes in seawater chemistry lead to measurable reactions of marine species and ecosystems, and are known to have negative repercussions for many calcifying organisms [4–7], including certain scleractinian corals, bryozoans, molluscs, and echinoderms. Decreased carbonate ion concentration $[CO_3^{2-}]$ can disrupt the physiologically regulated biomineralization mechanism that generates, preserves and maintains calcium carbonate ($CaCO_3$) structures, e.g., exoskeleton, test, spine, tube feet, teeth, pedicellariae and spicules [4,8–10]. This mechanism involves mineral formation, characteristics, morphogenesis, and organic molecules [9,11–13]. This is particularly the case for organisms precipitating a skeleton composed of high Mg-calcite (HMC) i.e., with a significant concentration of magnesium carbonate ($MgCO_3$) in the carbonate (>4 mol% $MgCO_3$) [14,15]. HMC is the most soluble of the polymorphs of crystalline $CaCO_3$ and is thermodynamically metastable [14,16]. Furthermore, previous studies have shown that OA significantly affects the size and weight of shells or skeletons of many marine calcifiers [17–19]. In contrast, some calcifiers, mainly photosynthesizing or photosymbiotic ones, including some corals and algae, show positive responses in calcification and growth values as they benefit from increased CO_2 concentrations by an enhanced photosynthesis rate, which provides additional potential energy for the calcification process [20].

At the same time, seawater temperature (ocean warming, OW) influences the ecophysiology of marine organisms [21], skeletal mineralogy [11,14,22–27], growth rate [28], and mineral growth control mechanisms [29]. Previous studies have found inconsistent reactions to increased temperature exposure, ranging from no effect to significant changes in skeletal mineralogy. For example, the skeletal Mg/Ca ratios of the scleractinian coral *Acropora* sp. [30], the sea urchin *Paracentrotus lividus* [25], the foraminifera *Planoglabratella operculari, Quinqueloculina yabei* [31] and *Ammonia tepida* [32] increase with temperature; conversely, the skeletal Mg/Ca ratios of the sea urchin *Lytechinus variegatus* was significantly lower in individuals kept at high temperature (~30 °C) compared to the ambient temperature (~26 °C) [26]. Skeletal Sr/Ca ratios decreases with increasing temperature in the scleractinian coral *Acropora* sp. [30,33], Sr/Ca ratios increases with increasing temperature in the foraminifera *A. tepida* [32] or Globigerinoides ruber [34], and is not significantly affected by temperature in the foraminifera *Trifarina angulosa* [35]. However, the effects of temperature on skeletal mineralogy and structure are still largely unclear and are likely influenced by phylogenetic factors, growth rate [22,23,36], latitude [24], biological 'vital effects' [37], and stage or species-specific differences [38].

The effects of combined OA and OW on marine calcifiers organisms are thought to be additive, antagonistic, or synergistic [39]. Previous meta-studies observed complex responses of calcifying marine organisms to combined OW and OA that vary from a significant negative effect to no effect on the organism [40]. Moreover, there is a trend toward enhanced sensitivity (i.e., the capability to sense and respond to environmental alterations) of biomineralization, growth, survival, and life stages development to OA in corals, echinoderm, and molluscs when being concurrently exposed to elevated temperatures [41]. Besides, there is growing evidence that elevated temperatures can exacerbate microstructure disruption caused by elevated pCO_2 in ectotherm species, e.g., in some molluscs such as the giant clam *Tridacna maxima* [42] and the mussel *Mytilus edulis* [43,44], and echinoderms such as the sea urchin *Tripneustes gratilla* [45]. However, these synergistic effects of OA and OW seem to be complex, and the magnitude of effect sizes and organism response varies between taxa groups, trophic level, habitat and life stages [40,41,46].

Echinoderms comprise a wide variety of taxa, with a complex calcium carbonate biomineralization. Their skeleton is formed within the syncytium by progressive crystallization of a transient amorphous calcium carbonate phase (ACC) [47], through a biologically controlled intracellular mechanism within vesicles or vacuoles formed by fused cell membranes inside cells [11,22,48]. The echinoderm endoskeleton is composed of a complex three-dimensional porous microstructure (stereom) with connecting trabeculae (i.e., mesh-like interconnecting matrix rod of calcite skeleton) [11], which are composed of 99.8–99.9% weight/weight (*w*/*w*) HMC with Sr as the primary trace element [49,50], and the other 0.1–0.2% (*w*/*w*) of the skeleton being organic components consisting of proteins and glycoproteins, called the intrastereomic organic matrix (IOM) [51,52]. The IOM has critical functions in the biomineralization process during the transient ACC phase by stabilizing the skeleton, controlling mineral incorporation into the skeleton, and controlling the nucleation and morphology of the skeletal crystals [53,54]. The trace element concentration of biogenic $CaCO_3$ is affected by biological factors, e.g., phylogeny, life stage, food supply, as well as by physical-chemical factors, e.g., temperature, salinity, seawater carbonate chemistry, concentration of Mg^{2+} and Ca^{2+} ions in the seawater, light, and hydrostatic pressure [14,27,37,49,55,56]. Seawater chemistry influences the mineralogy of the echinoderm skeleton by changing the physiological cost of sustaining the biological control of intracellular chemistry [57]. For echinoderms with their skeletons being composed of HMC ($MgCO_3$ between 2.5% and 39% [8]), their skeletons become more soluble under OA conditions [16,50,58,59].

To gain a better understanding of the effects of combined OA and OW on biomineralization, systematic comparative studies across a phylogenetically diverse spectrum of taxa are needed [57]. The present study contributes to this goal by investigating the mineral composition and skeleton microstructure of *Aquilonastra yairi* (phylum Echinodermata, class Asteroidea, family Asterinidae) under controlled OA and OW conditions. This asterinid starfish thrives in the marine intertidal and coral reefs of the Red Sea and the Mediterranean Sea, where it lives in crevices and beneath corals or rocks [60]. It has an important ecological function as a grazer of marine algae, bacterial mats, detritus, and other fragments of food [61]. Previous studies have indicated deleterious effects of combined OA and OW on the physiological performances of asterinid starfish [62–65], while the effects of each stressor and their potential synergetic effects on skeletal mineral ratio and microstructure are currently still poorly understood.

2. Materials and Methods

2.1. Experimental Design and Control of Seawater Chemistry

In this study, starfish *A. yairi* was used as a model organism to investigate the combined effects of OW and OA. A total of 342 specimens of *A. yairi* (size 3–11 mm) from the cultivated stock of the MAREE (Marine Experimental Ecology facility) of ZMT, Bremen, Germany, were studied in the present experiment. Following acclimation procedures (see electronic Supplementary Materials), the starfish were cultured for 90-days in six different treatments, namely at two different temperatures (ambient temperature: 27 °C, and high temperature: 32 °C) crossed with three levels of pCO_2 (low pCO_2: 455 µatm, medium pCO_2: 1052 µatm, and high pCO_2: 2066 µatm). All six treatments were replicated in three aquaria (electronic Supplementary Materials Figure S1).

Target temperature and pCO_2 levels were ramped up gradually over the first ten days to avoid physiological shock. Then the temperature (mean ± SE) in the treatment tanks was maintained at 27 ± 0.05 °C and 32 ± 0.08 °C, respectively, using a closed circle heating system (Heaters Titanium Tube 600 W, Schego Schemel & Goetz, Offenbach, Germany), controlled with a programmable thermostat. The mixture of the gas bubbled into the seawater in the bottom storage compartment sump was reached by blending compressed CO_2-free air and compressed CO_2 (pure CO_2 provided by Linde GmbH, Pullach, Germany) using electronic solenoid-valve mass-flow controllers (HTK Hamburg GmbH, Hamburg, Germany) in accordance with the standard operating procedure (SOP) for ocean CO_2 measurements [66]. Details of the seawater chemistry control and manipulation are provided in the electronic Supplementary Materials. The seawater parameters and carbonate chemistry for the experimental exposures are given in Table 1.

Table 1. Seawater chemistry values measured during a 90-day experimental period for *A. yairi* reared under two temperature levels (27 °C and 32 °C) crossed with three levels of pCO_2 (455 µatm, 1052 µatm, and 2066 µatm). A_T, total alkalinity; DIC, dissolved inorganic carbon; pCO_2, partial pressure of CO_2; $[CO_3^{2-}]$, carbonate ion concentration; $[HCO_3^-]$, bicarbonate ion concentration; $[CO_2]$, dissolved CO_2; Ω_{Ca}, calcite saturation state; Ω_{Ar}, aragonite saturation state. Data are presented as mean values ± SE.

Treatment	Measured Parameters					
	Salinity (PSU)	Temperature (°C)	$pH_{(NBS\ scale)}$	$pH_{(total\ scale)}$	A_T (µmol/kg-SW)	DIC (µmol/kg-SW
27 °C: 455 µatm	34.56 ± 0.12	27.48 ± 0.06	8.13 ± 0.00	8.00 ± 0.00	2504.42 ± 15.33	2168.86 ± 15.23
27 °C: 1052 µatm	34.73 ± 0.06	27.23 ± 0.04	7.87 ± 0.01	7.74 ± 0.01	2514.45 ± 16.78	2340.99 ± 11.62
27 °C: 2066 µatm	34.75 ± 0.05	27.34 ± 0.03	7.60 ± 0.01	7.47 ± 0.01	2539.83 ± 38.86	2479.40 ± 36.58
32 °C: 455 µatm	34.65 ± 0.08	32.03 ± 0.05	8.13 ± 0.00	8.00 ± 0.00	2510.27 ± 47.18	2134.38 ± 38.40
32 °C: 1052 µatm	34.78 ± 0.05	32.10 ± 0.04	7.87 ± 0.00	7.74 ± 0.00	2532.19 ± 40.05	2325.20 ± 40.02
32 °C: 2066 µatm	34.76 ± 0.02	32.20 ± 0.08	7.60 ± 0.01	7.47 ± 0.01	2584.38 ± 41.93	2493.30 ± 37.87

Treatment	Calculated Parameters					
	pCO_2 (µatm)	$[CO_3^{2-}]$ (µmol/kg-SW)	$[HCO_3^-]$ (µmol/kg-SW)	$[CO_2]$ (µmol/kg-SW)	Ω_{Ca}	Ω_{Ar}
27 °C: 455 µatm	456.13 ± 8.24	245.64 ± 2.52	1911.05 ± 14.94	12.17 ± 0.22	5.96 ± 0.06	3.96 ± 0.04
27 °C: 1052 µatm	1059.58 ± 32.04	138.12 ± 4.64	2178.74 ± 9.58	28.42 ± 0.86	3.35 ± 0.11	2.22 ± 0.07
27 °C: 2066 µatm	2075.40 ± 30.99	81.18 ± 2.35	2342.71 ± 34.29	55.51 ± 0.82	1.97 ± 0.06	1.31 ± 0.04
32 °C: 455 µatm	453.78 ± 6.51	273.82 ± 8.23	1849.66 ± 30.77	10.90 ± 0.16	6.71 ± 0.20	4.52 ± 0.14
32 °C: 1052 µatm	1045.15 ± 44.00	162.68 ± 4.67	2147.47 ± 38.59	25.05 ± 1.06	3.98 ± 0.11	2.69 ± 0.08
32 °C: 2066 µatm	2057.31 ± 74.42	99.46 ± 4.64	2344.63 ± 34.94	49.20 ± 1.78	2.44 ± 0.11	1.64 ± 0.08

2.2. Skeletal Mineral Composition Analysis

On days 45 and 90 of the experimental treatment, six specimens from each replicate tank (i.e., 36 in total on each of those days) were randomly collected and rinsed in Milli-Q (18.2 MΩ) water before drying for ~48 h at 40 °C. In preparation for trace element analysis, to remove organic material from the skeletal matrix, dried starfish were soaked in hydrogen peroxide (H_2O_2) [67–69] for 24 h and subsequently cleaned mechanically, i.e., residual organic material was removed by forceps, and further potential contaminations were removed with deionized water in an ultrasonic bath. Then, the sample material was manually ground using a mortar and pestle. The powdered samples were kept at room temperature in sealed vials until analysis.

The element concentration of Ca, Mg and Sr in the skeleton was determined with a Spectro CIROS Vision (SPECTRO Analytical Instruments GmbH, Kleve, Germany) inductively coupled plasma optical emission spectroscope (ICP-OES). The samples (weighing 0.02–0.1 mg) were digested with concentrated nitric acid (HNO_3) and H_2O_2 (high-purity of trace metal grade reagent). The solutions were then diluted to the acidity of 0.5 M HNO_3 with aliquots of 0.1 mL and weighed again. Instrument calibration solutions (Inorganic Ventures^{TM} 1000 ppm standard stock solution) were prepared using single-element standards in proportion to the *A. yairi* skeleton concentrations. Measurements of all starfish samples were done routinely against the international reference standard JLs^{-1}, a coral in-house working standard ($ZMT\text{-}CM_1$), and HNO_3 blanks. Mg and Sr mineral elements are reported as a ratio over calcium (Ca), i.e., Mg/Ca, and Sr/Ca and over total skeletal material, i.e., Ca_{norm}, Mg_{norm}, and Sr_{norm}, to account for minor organic material still present on or within the carbonate skeleton.

2.3. Analysis of the Skeleton Microstructure

One specimen from the 45-day and 90-day incubations from each treatment tank was randomly selected (n = 12), washed, and prepared for the SEM analysis. Each starfish was cut with dissecting scissors around the part of arms and cleaned of soft tissue. Organic material was dissolved using a 30% H_2O_2 solution buffered in NaOH (0.1 N) at room temperature for 24 h. Skeletons were then rinsed with distilled water repeatedly to remove any remaining organic material and then air-dried for 48 h at room temperature. Skeleton plates were then mounted on a stub with carbon based tape and gold-sputtered (Cressington Sputter Coater 108 auto, Cressington Scientific Instruments, Watford, UK) for 30 s. Secondary electron images (SE) were generated with a scanning electron microscope (SEM; Tescan Vega3 XMU, Brno-Kohoutovice, Czech Republic) to characterize the skeleton microstructure, using a beam voltage of 5 kV for a magnification of up to 3000×. All SEM micrographs were examined for any visible differences between treatments, including signs of dissolution, surface smoothness, the shape of stereom pores, and the shape of inner matrix aperture pores.

2.4. Statistical Analysis

Statistical analysis was performed using the software R, version 4.1.3 [70]. Normality of data distribution and homogeneity of variance was tested with the Shapiro–Wilk statistic W test (α = 0.05) [71] and Levene's test (α = 0.05) [72], respectively, and indicated that all data of skeletal mineral ratios were normally distributed and the homoscedasticity assumption for the data was equal. The effects of temperature, pCO_2, incubation time and their interactions on skeletal mineral element to calcium ratios (Mg/Ca and Sr/Ca) and skeletal mineral element to total skeletal material ratios (Ca_{norm}, Mg_{norm}, and Sr_{norm}) were examined using three-way analysis of variance (ANOVA), and Tukey HSD post hoc analyses were conducted using agricolae R-package 1.3-5 [73]. Temperature, pCO_2, and incubation time were fixed factors, while skeletal mineral ratios were used as response variables. All statistics were evaluated with a significance level of α = 0.05.

3. Results

Over the duration of the experiment (90 days), the starfish mortality rate was low and only found in the high-temperature treatment. In general, the results indicate that elevated temperature and pCO_2 changed the skeletal mineral composition, whereas elevated pCO_2 affected skeletal microstructure in *A. yairi*.

3.1. Elemental Composition of Skeletal Carbonate

Overall, a relatively small range of Mg/Ca ratio values were observed across all our treatments (181.95–204.26 mmol/mol). The starfish had consistently higher Mg/Ca ratios in the 32 °C treatments (190.90 ± 1.41 mmol/mol, mean ± SE) than those held at 27 °C (187.59 ± 0.83 mmol/mol, mean ± SE) throughout all pCO_2 concentration levels. Both incubation time and temperature had a main effect on skeletal Mg/Ca ratio (p = 0.049 and p = 0.033, respectively; Table 2). Inter-individual variability in skeletal Mg/Ca ratios was substantially higher in starfish subjected to high temperatures (32 °C) compared to those exposed to ambient temperatures (27 °C) in all pCO_2 combined treatments (Figure 1A). No consistent pCO_2 effect as the sole factor was found, and the Mg/Ca ratio displayed the typical parabolic responses to pCO_2 (Figure 1A). Elevated pCO_2 as the sole stressor did not significantly affect skeletal Mg/Ca ratios (p = 0.414, Table 2). The interaction of temperature: pCO_2: incubation time on starfish skeletal Mg/Ca ratios was significant (p = 0.014, Table 2). However, Tukey's HSD post hoc analysis did not reveal any significant interactions in Mg/Ca ratios (Table 2 and electronic Supplementary Materials Table S1).

Table 2. Summary of three-way ANOVA results for the skeletal mineral ratios of *A. yairi* exposed to temperature (27 °C, 32 °C) crossed with elevated pCO_2 (455 µatm, 1052 µatm, 2066 µatm) treatments for 45 and 90 days incubation time. Tukey HSD post hoc tests were performed where ANOVA results indicated significant effects of one or several factors (incubation time, pCO_2, temperature) with p-values adjusted for multiple testing (p_{adj}). Bold terms indicate a significant difference ($p < 0.05$).

Skeletal Mineral Ratio	df	F	Pr (<F)	Post-Hoc Test Result
Mg/Ca ratio				
Incubation time	1	4.319	**0.049**	90 days > 45 days
pCO_2	2	0.915	0.414	
Temperature	1	5.144	**0.033**	32 °C > 27 °C
Incubation time: pCO_2	2	0.715	0.500	
Incubation time:temperature	1	0.039	0.845	
pCO_2: temperature	2	0.478	0.626	
Incubation time: pCO_2: temperature	2	5.143	**0.014**	n.s., electronic Supplementary Materials Table S1
Residuals	24			
Sr/Ca ratio				
Incubation time	1	9.027	**0.006**	90 days > 45 days
pCO_2	2	0.405	0.671	
Temperature	1	0.481	0.495	
Incubation time: pCO_2	2	1.794	0.188	
Incubation time: temperature	1	1.519	0.230	
pCO_2: temperature	2	2.016	0.155	
Incubation time: pCO_2: temperature	2	1.132	0.339	
Residuals	24			
Ca_{norm} ratio				
Incubation time	1	4.951	**0.036**	90 days > 45 days
pCO_2	2	1.242	0.307	
Temperature	1	0.043	0.838	
Incubation time: pCO_2	2	1.711	0.202	
Incubation time: temperature	1	0.59	0.450	
pCO_2: temperature	2	1.471	0.250	
Incubation time: pCO_2: temperature	2	1.65	0.213	
Residuals	24			
Mg_{norm} ratio				
Incubation time	1	24.523	**0.001**	90 days > 45 days
pCO_2	2	0.582	0.566	
Temperature	1	4.206	**0.051**	32 °C > 27 °C
Incubation time: pCO_2	2	0.616	0.548	
Incubation time: temperature	1	0.621	0.438	
pCO_2: temperature	2	0.609	0.552	
Incubation time: pCO_2: temperature	2	2.133	0.140	
Residuals	24			
Sr_{norm} ratio				
Incubation time	1	9.814	**0.005**	90 days > 45 days
pCO_2	2	0.500	0.613	
Temperature	1	0.156	0.696	
Incubation time: pCO_2	2	1.862	0.177	
Incubation time: temperature	1	1.405	0.248	
pCO_2: temperature	2	1.655	0.212	
Incubation time: pCO_2: temperature	2	0.707	0.503	
Residuals	24			

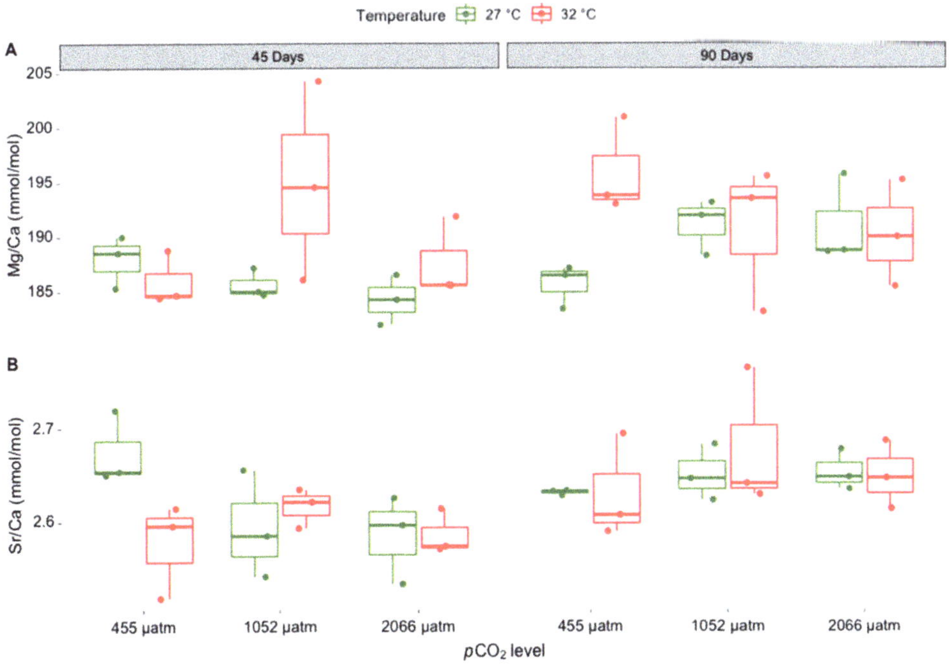

Figure 1. Changes in skeletal properties (**A**) Mg/Ca ratios (mmol/mol) and (**B**) Sr/Ca ratios (mmol/mol) of skeletal carbonate in *A. yairi* exposed to different temperatures (27 °C and 32 °C) crossed with different pCO_2 concentrations (455 µatm, 1052 µatm, and 2066 µatm) measured after 45 and 90 days of incubation ($n = 36$). Dots represent individual skeletal mineral values, displayed with jitter to avoid overlap.

Sr/Ca ratios ranged from 2.52 mmol/mol to 2.76 mmol/mol across treatments (Figure 1B). The Sr/Ca ratio at 32 °C had the highest fluctuation in values compared to 27 °C, where Sr/Ca ratios at 27 °C treatments (2.63 ± 0.01 mmol/mol, mean ± SE) were slightly higher than for the 32 °C treatments (2.62 ± 0.01 mmol/mol, mean ± SE) (Figure 1B). Inter-individual variability of Sr/Ca ratios were substantially higher for medium and high pCO_2 treatments (1052 µatm, 2066 µatm) compared to low pCO_2 (455 µatm) treatments; this was the case at both temperature levels (Figure 1B). No significant response of Sr/Ca ratio to differences in the pCO_2 and temperature as combined or as sole stressors (Table 2). However, skeletal Sr/Ca ratios were significantly altered over incubation time ($p = 0.006$, Table 2), with increasing values from samples taken after 45 days compared to those collected after 90 days.

Ca_{norm} ratios showed relatively variable values, which ranged between 647.74 mg/g and 755.30 mg/g across all treatments (Figure 2A). Ca_{norm} ratios were changed over the incubation time ($p = 0.036$, Table 2). The Ca_{norm} ratios at 27 °C rose from 699.73 ± 5.41 mg/g (mean ± SE) at 45 days to 709.02 ± 4.33 mg/g (mean ± SE) at 90 days incubation time (enhanced 1.33%), whereas the Ca_{norm} ratios at 32 °C rose from 693.51 ± 8.93 mg/g (mean ± SE) at 45 days to 712.60 ± 7.48 mg/g (mean ± SE) at 90 days incubation time (enhanced 2.74%). Ca_{norm} ratios were not significantly affected by temperature ($p = 0.838$, Table 2) or pCO_2 ($p = 0.307$, Table 2) as a single factor, nor as a combined factor ($p = 0.250$, Table 2).

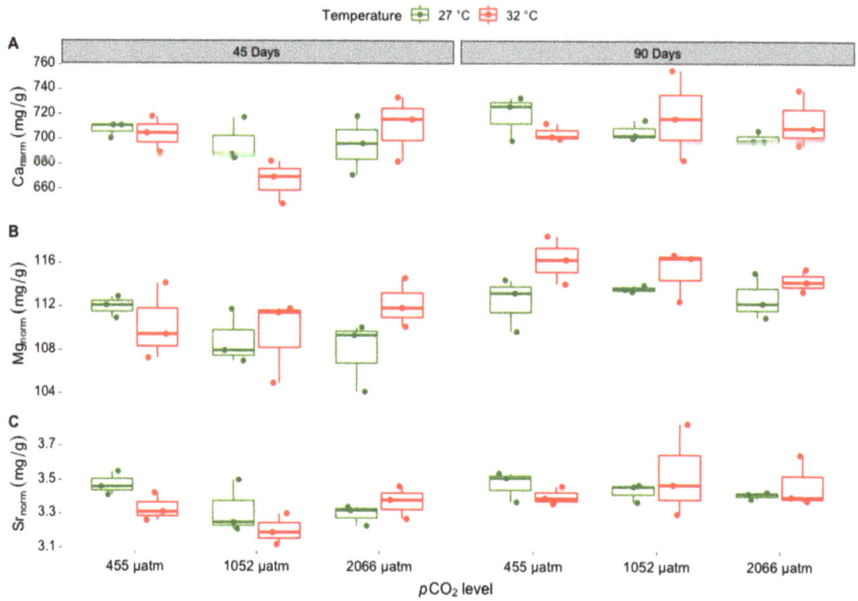

Figure 2. Ratios between skeletal mineral element to total skeletal material (**A**) Ca_{norm} ratios (mg/g), (**B**) Mg_{norm} ratios (mg/g), and (**C**) Sr_{norm} ratios (mg/g) in *A. yairi* exposed to elevated temperatures levels (27 °C and 32 °C) crossed with increased pCO_2 concentrations (455 µatm, 1052 µatm and 2066 µatm) measured after 45 and 90 days of incubation (n = 36). Dots represent individual skeletal mineral values, displayed with jitter to avoid overlap.

Mg_{norm} ratios from the skeleton of *A. yairi* were significantly altered by incubation time (p = 0.001, Table 2). Mg_{norm} ratios increased from 45 days to 90 days in both temperature treatment conditions (Figure 2B). At 27 °C, the Mg_{norm} ratios increased from 109.57 ± 0.94 mg/g (mean ± SE) at 45 days to 112.98 ± 0.57 mg/g (mean ± SE) at 90 days incubation time (i.e., 3.11% increase), while at 32 °C, the Mg_{norm} ratios increased from 110.61 ± 1.04 mg/g (mean ± SE) at 45 days to 115.31 ± 0.63 mg/g (mean ± SE) at 90 days incubation time (4.25% increase). There was no significant effect of pCO_2 nor any combined effect of the factors (Table 2). However, temperature led to a marginal increase in the Mg_{norm} ratios (p = 0.051, Table 2).

Sr_{norm} ratios were altered over incubation time (p = 0.005, Table 2) at both temperature treatments (Figure 2B). The mean value of the Sr_{norm} ratio at 27 °C increased by 2.38% from day 45 (3.36 ± 0.04 mg/g, mean ± SE) to day 90 (3.44 ± 0.02 mg/g, mean ± SD). Similarly, the mean value of Sr_{norm} ratio at 32 °C increased from 45 days (mean ± SD, 3.30 ± 0.04 mg/g) to 90 days (3.47 ± 0.06 mg/g, mean ± SD) incubation time (5.15% increase). However, there was no significant effect of combined stressor factors nor solely stressor factors on Sr_{norm} ratios (Table 2).

3.2. Skeletal Microstructure

High magnification SEM micrographs showed marked differences in skeletal structure between low pCO_2 (455 µatm) compared to medium and high pCO_2 treatments (1052 µatm, 2066 µatm, respectively) at both ambient and high temperatures (27 °C, 32 °C, respectively). The skeletal structure in low pCO_2 crossed with ambient (27 °C: 455 µatm) and high temperatures (32 °C: 455 µatm) revealed no remarkable differences between 45-day and 90-day incubation time. The stereom pores were arranged equally in shape and the aperture pores of the inner matrix were relatively equal in shape, while the trabecular surface was smooth (Figure 2A,B,G,H and Table 3).

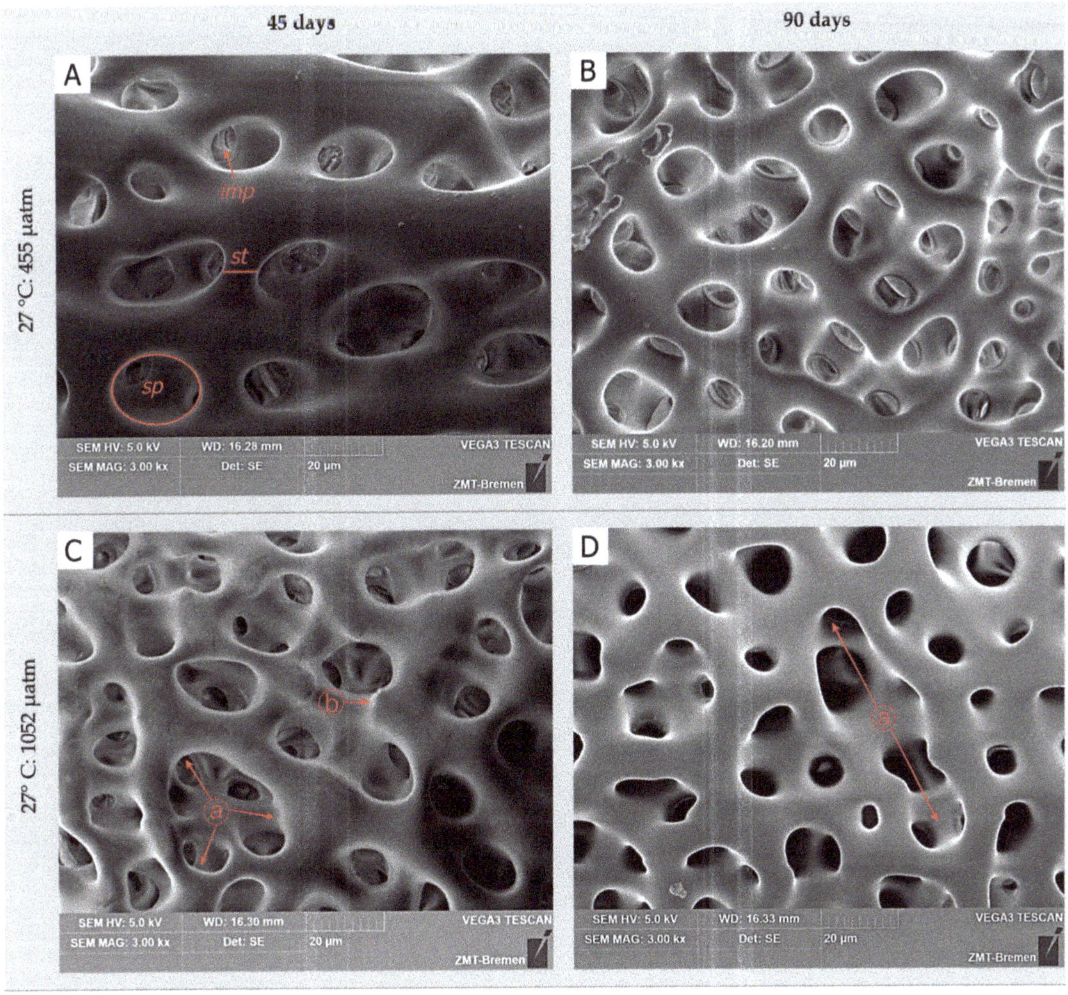

Figure 2. *Cont.*

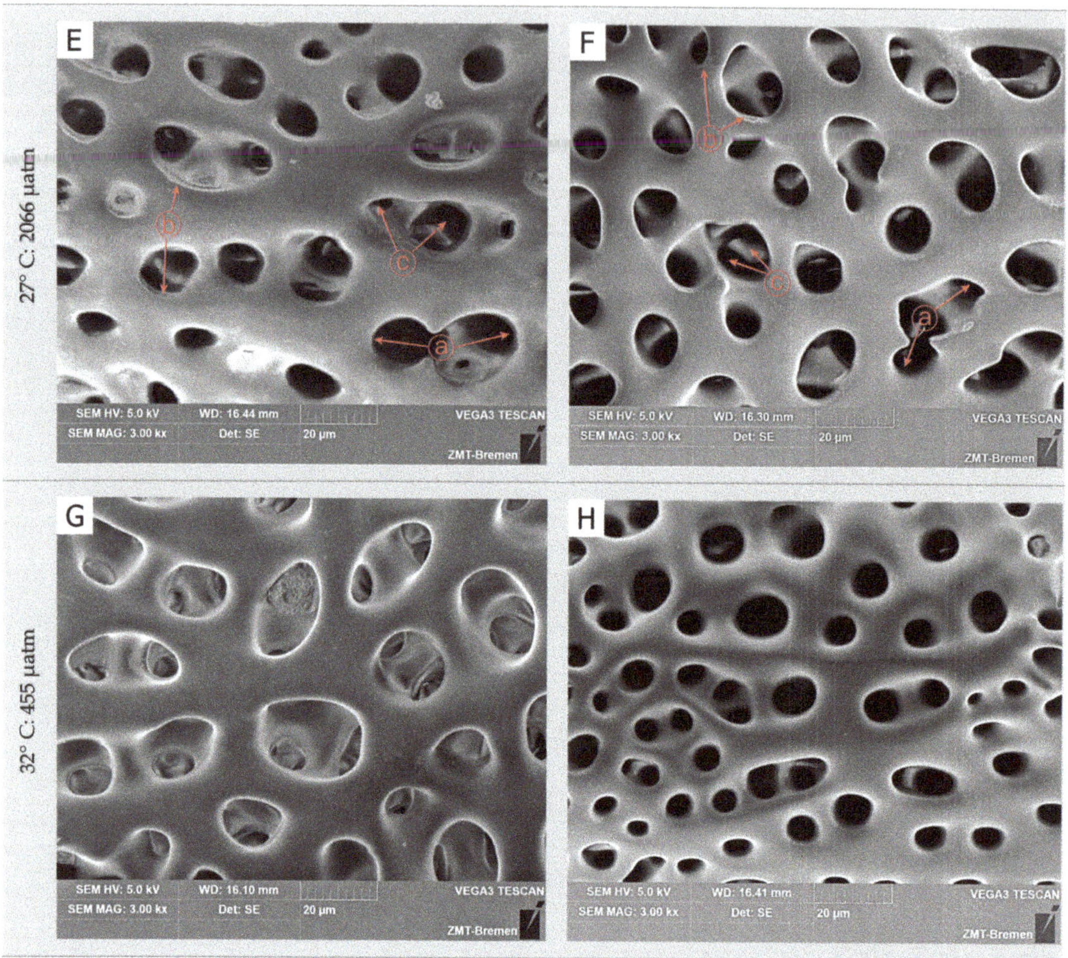

Figure 2. *Cont.*

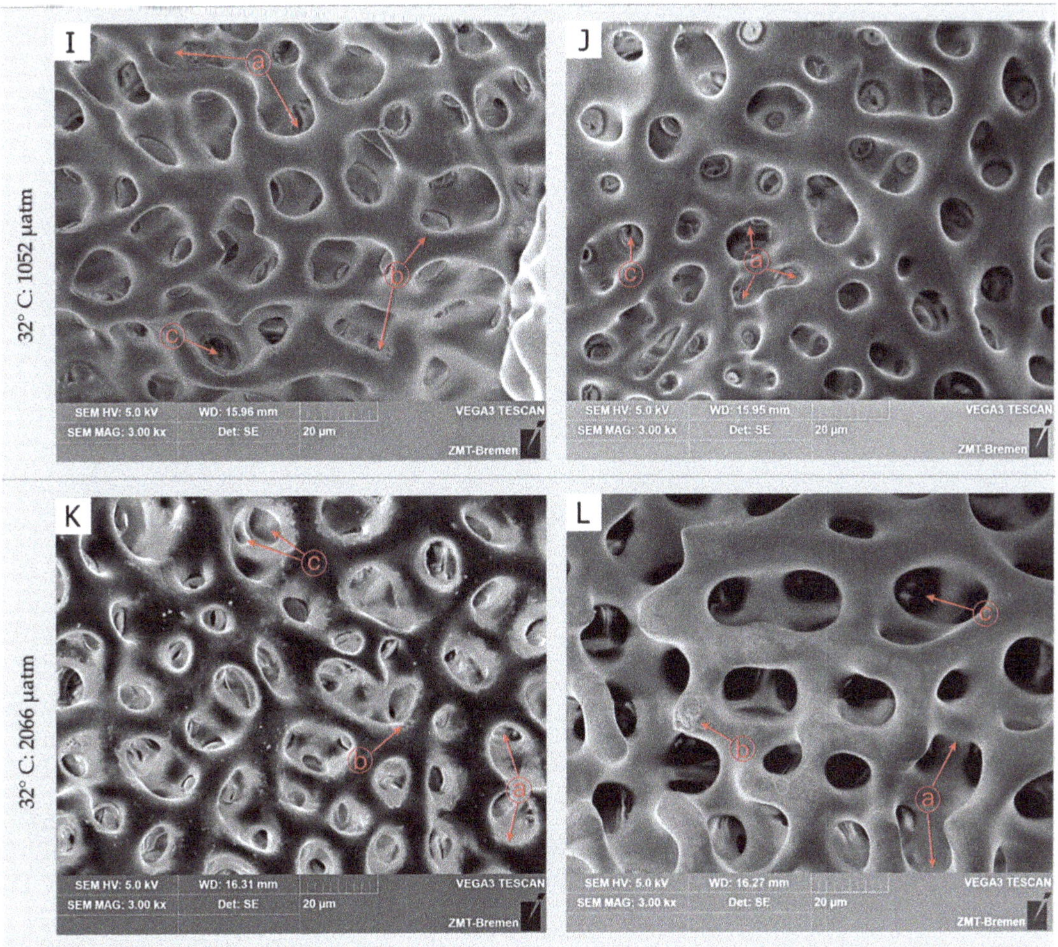

Figure 2. SEM micrographs of the skeletal microstructure of *A. yairi* cultured under two temperature levels (27 °C and 32 °C) crossed with three pCO_2 concentrations (455 µatm, 1052 µatm and 2066 µatm). Stereom microstructure after 45 (**A**,**C**,**E**,**G**,**I**,**K**) and 90 days (**B**,**D**,**F**,**H**,**J**,**L**) of incubation time. *sp*: stereom pore; *st*: skeleton trabeculae; *imp*: stereom inner matrix pore; (a) the galleries stereom pores are less-equal in shape; (b) dissolution in calcium carbonate skeleton; (c) increased size of inner matrix aperture pores. Technical image acquisition, SEM mode: SE, SEM HV: 5.0 kV, SEM magnification: 3000×.

In contrast, after 45-day and 90-day incubation times, the skeleton from medium and high pCO_2 treatments at ambient temperatures showed stereom structures that were more variable in shape compared to the control treatment, and signs of degradation, i.e., dissolution, were observed on the surface of the trabeculae (Figure 2C–F and Table 3). Furthermore, under high temperatures, these medium and high pCO_2 treatments in addition result in signs of skeletal degradation observed at the trabeculae surface, while the apertures of the inner matrix pores were wider (i.e., un-equal in shape) compared to the control treatment and the ambient temperature crossed with medium and high pCO_2 treatments (Figure 2I–L and Table 3).

Table 3. Skeletal microstructure characteristics of *A. yairi* under crossed temperatures and pCO_2 conditions at different incubation times as observed with scanning electron microscopy (SEM). ND: the skeleton surface is smooth and has no signs of degradation; DS: the skeleton surface had degradation signs; HD: the skeleton surface had high degradation signs; ES: the stereom pores were equal in shape; US: the stereom pores were un-equal in shape; HU: the stereom pores were highly un-equal in shape; EP: the inner matrix aperture pores were relatively equal in shape; UP: the inner matrix aperture pores were relatively un-equal in shape (wider). (n = 12).

Incubation Time	Temperature	pCO_2	Skeletal Surface	Stereom Pores	Inner Matrix Pores
45 days	27 °C	455 µatm	ND	ES	EP
		1052 µatm	DS	US	EP
		2066 µatm	DS	US	UP
	32 °C	455 µatm	ND	ES	EP
		1052 µatm	DS	US	UP
		2066 µatm	HD	HU	UP
90 days	27 °C	455 µatm	ND	ES	EP
		1052 µatm	DS	US	EP
		2066 µatm	DS	US	UP
	32 °C	455 µatm	ND	ES	EP
		1052 µatm	DS	US	UP
		2066 µatm	HD	HU	UP

4. Discussion

Our results imply that temperature plays a primary regulatory role in Mg concentration in the skeletal carbonate of *A. yairi*, where the Mg/Ca ratios increase under high-temperatures. This corroborates results from previous studies that found higher Mg concentrations associated with higher temperatures in echinoid and asteroid species [23,24,74]. This temperature association was previously attributed to kinetic factors affecting ion discrimination [23] and also to the physiological mechanisms that control Mg absorption in cells [75] during the biomineralization process. The increased Mg content of HMC was connected to an amorphous calcium magnesium carbonate (ACMC) precursor [76,77]. At high temperatures, the aqueous Mg^{2+} solvation energy barrier becomes lower [78]; hence this condition might favor more Mg^{2+} to be incorporated into the calcite lattice, encouraging the formation of ACMC, which later transforms into HMC.

Echinoderms are generally considered relatively poor regulators of internal acid–base balance, where the range of regulatory capacities is species-specific [79]. In hypercapnic conditions (high pCO_2, low pH), echinoderms increase the bicarbonate concentration in their coelomic fluid and practice passive skeletal dissolution to support internal acid–base regulatory functions due to acidosis [80]. Higher Mg concentrations in the skeleton in conjunction with a degradation in the inner skeleton (see Section 3.2) due to skeletal dissolution, as documented in the current studies, support the assumption of a trade-off mechanism [65,81]. The released HMC mineral is then used as an active buffering mechanism to compensate for changes in internal pH that may help to avoid or reduce physiological impacts.

We noticed that no single (temperature or pCO_2) nor combined stressor treatment affected skeletal Sr (Sr/Ca and Sr_{norm}) in *A. yairi*. This contrasts with previous studies on the starfish *Asterias rubens* [49] and the sea urchin *Paracentrotus lividus* [25], which reported that the Sr/Ca ratio depended on temperature. These contrasting results indicate a species-specific skeletal Sr control mechanism in the echinoderm group that may respond to the stressors. The calcification rates might play an important role in skeletal Sr precipitation rather than direct dependence on temperature. It is difficult to discriminate the effects of temperature and pCO_2 on skeletal Sr. Our data indicated that the skeletal Sr was increased over incubation time in all treatment combinations (electronic Supplementary Materials Figure S2) except the 27 °C: 455 µatm treatment, assuming that temperature and pCO_2 might have an indirect effect on skeletal Sr through their control over the calcification process [23,34,69,82], which are common in biogenic calcites [83,84]. However, the incorporation pathways of Sr into the echinoderm ACC are still poorly understood, so further studies are needed.

We found a strong correlation between skeletal Sr and skeletal Mg ($R^2 = 0.04$, $p < 0.001$, electronic Supplementary Materials Figure S3), as Sr_{norm} increases with increasing Mg_{norm} ratio. Hence, Mg might play a role in affecting the Sr precipitation. Sr ions cannot easily substitute for Ca ions due to differences in cations ions size and weight (Sr^{2+} ionic radius of 1.18 Å: Ca^{2+} ionic radius of 1.00 Å [85]), which makes them incompatible in calcite [86]. High concentrations of Mg^{2+} (ionic radius of 0.72 Å [85]) that are incorporated in inorganic calcite can distort the crystal lattice (i.e., lattice deformation), which increases the size of calcium lattice positions and thus allows for increased incorporation of Sr^{2+} [87].

Similar to skeletal Sr, we found no direct influence of temperature or pCO_2 on the Ca_{norm} ratio (Table 2). However, the highest inter-individual variability of Ca_{norm} indicates that temperature and/or pCO_2 considerably alter the production of skeletal Ca in *A. yairi*, where Ca_{norm} ratios exhibited an increase with incubation time, except for the treatment of 27 °C: 455 µatm (~703 mg/g) where the ratio was relatively unchanged (electronic Supplementary Materials Figure S4). $CaCO_3$ is more soluble at lower temperatures [88] and high pCO_2 [20]. Under OW conditions, starfish seem to be able to boost their capacity to control calcification through the modulation of the intracellular calcifying fluids pH to produce $CaCO_3$. This might provide them with higher resistance and resilience towards the effects of OA. The increased energetic costs of this physiological response are indicated by elevated respiration rates [65]. Hence, in the long term, this phenotypic plasticity and biocalcification control mechanism might have further physiological implications for the starfish.

The hypercapnic conditions of the high pCO_2 treatment in our experiment have resulted in visible adverse effects on the skeletal structures in both ambient as well as elevated temperatures (Figure 2, Table 3). We detected changes in the stereom pores, the inner matrix pores and degradation of the skeleton surface through reduced deposition or dissolution, which increased over incubation time in the starfish that were exposed to medium and high pCO_2. These skeletal alterations might be due to the thin layer of epithelial tissue covering the starfish body wall [89], which acts as a protective membrane between the skeleton and the seawater [90]. Furthermore, the degradation of the epithelial tissue could also explain why the combination of OW with OA enhances the degradation of the skeletal structure that was observed in the high pCO_2 treatment (32 °C: 2066 µatm). We hypothesize that the epithelial tissue might become weaker in its function due to degradation of the epithelium cells during long-term experiments. Epithelial tightness is controlled by a protein series that molds a seal among epithelial cells [90], which are sensitive to elevated temperature.

The seawater chemistry factor, which was underlying the altered skeletal structure in *A. yairi* might involve calcium carbonate saturation state (Ω). Skeletal degradation provides evidence that the skeleton of *A. yairi* was negatively affected by lowered calcite saturation state (Ω_{ca}) as a function of increasing pCO_2 (Table 1), which directly affects the biomineralization and dissolution of the $CaCO_3$ in skeletal structures [17,91]. Near Ω_{ca} ~3, we found stereom alteration and degradation signs in trabeculae, which indicates that *A. yairi* was facing difficulty in producing and maintaining their skeleton compared to low pCO_2 treatments ($\Omega_{ca} > 5$). Previous studies found that the skeletal morphometric development of echinoid (e.g., sea urchin *Lytechinus variegatus*) was significantly affected by Ω_{ca} 2.53 [92]. Furthermore, increased susceptibility of the skeleton to dissolution through OA was suggested to be due to the skeleton composed of HMC, which is 30 times more soluble than pure calcite [47,93]. When the qualitative approach depicts changes in the skeleton micro-morphology because of exposure to elevated pCO_2 and temperature (Table 3, Figure 2), further investigation using quantitative analysis (e.g., stereom and inner matrix pore size) is required to quantify the alteration magnitude in skeleton microstructure as the deleterious impact of OA and OW [94,95].

Under OW and OA scenarios, calcifying marine organisms are expected to face unfavorable conditions to produce and maintain their skeletal HMC in order to sustain their biomechanical functions. Phenotypic plasticity requires re-allocation of energy as a trade-off and may represent a potential key mechanism for species viability [96]. For *A. yairi*, this mechanism might reflect strategies to maintain their skeletal HMC in a low pH environment by modulating their cellular chemistry to create isolated microenvironments of deposition, producing the specific mineral they necessitate, which involves significant energy re-allocation [97]. Since the asterinid starfish produces HMC, which is more soluble than calcite and aragonite, it signifies that more energy will be required to preserve the calcification process as the mechanism in both constructing and maintaining their skeletal components, leading to energy trade-offs against other physiological processes [65,98].

Changes in the shape of the stereom pores and degradation of the starfish skeleton potentially have implications for skeletal strength, stiffness, and function. Weakening of the structure possibly will reduce locomotion performance and result in a lower ability to resist predators and to face ocean currents, which then conveys consequences to the benthic community structure.

Supplementary Materials: The following supporting information can be downloaded at: https://www.mdpi.com/article/10.3390/jmse10081065/s1; Table S1: Tukey HSD post hoc test for the interactive effects of incubation time (45 and 90 days), pCO_2 (455 μatm, 1052 μatm, 2066 μatm) and temperature (27 °C, 32 °C) on skeletal Mg/Ca ratio of *A. yairi* using the 'agricolae R-package' for multiple comparisons to interrogate the main effects of incubation time, temperature and pCO_2 (incubation time: pCO_2: temperature = 0.014, Table 2); Figure S1: Schematic of ocean acidification and ocean warming experimental design with a fully factorial combination of low pCO_2 (455 μatm), moderate pCO_2 (1052 μatm), and high pCO_2 (2066 μatm) treatments with ambient temperature (27 °C) and high temperature (32 °C) treatments; Figure S2: Skeletal Sr values over incubation time in *A. yairi* exposed to elevated temperatures levels (27 °C and 32 °C) crossed with increased pCO_2 concentrations (455 μatm, 1052 μatm, and 2066 μatm). (**A**) Sr/Ca ratio (mmol/mol), (**B**) Sr_{norm} ratio (mg/g). Data were presented as mean (n = 36); Figure S3: Correlation of Sr_{norm} ratios (mg/g) against Mg_{norm} ratios (mg/g) (R^2 = 0.44, $F_{1,34}$ = 26.78, $p = 1.019 \times 10^{-5}$) in *A. yairi* exposed to elevated temperatures levels (27 °C and 32 °C) crossed with increased pCO_2 concentrations (455 μatm, 1052 μatm, and 2066 μatm). (n = 36); Figure S4: Skeletal Ca_{norm} (mg/g) ratio over incubation time in *A. yairi* exposed to elevated temperatures levels (27 °C and 32 °C) crossed with increased pCO_2 concentrations (455 μatm, 1052 μatm, and 2066 μatm). Data were presented as mean (n = 36). For more details, please see [66,99–106].

Author Contributions: Conceptualized and designed: M.K. and H.W.; methodology: M.K., S.S.D., M.S. and H.W.; investigation: M.K.; resources: M.K. and H.W.; sample processing and analysis: M.K.; data curation: M.K.; writing original draft preparation: M.K.; visualization: M.K.; writing-review and editing: M.K., S.S.D., M.S. and H.W.; supervision: H.W. All authors have read and agreed to the published version of the manuscript.

Funding: This research project was supported by the Academy Doctoral Research—Grant Leibniz Centre for Tropical Marine Research (ZMT), Germany and the Ministry of Education, Culture, Research, and Technology (MoECRT), Republic of Indonesia—Asian Development Bank (ADB) AKSI Project [grant number L3749-INO] to M.K.

Institutional Review Board Statement: Not applicable.

Informed Consent Statement: Not applicable.

Data Availability Statement: The datasets used and analyzed during the current study are present in the manuscript and/or supplementary materials. Data are available upon request from the corresponding author.

Acknowledgments: We would like to thank Andreas Kunzmann (ZMT, Bremen, Germany) for his advice on the starfish *A. yairi* as a model organism. Many thanks are due to the ZMT laboratory staff, namely Jule Mawick, Sebastian Flotow, Silvia Hardenberg, José Garcia, Nico Steinel, Matthias Birkicht and Fabian Hüge.

Conflicts of Interest: On behalf of all authors, the corresponding author states that there is no conflict of interest.

References

1. Watson, A.J.; Schuster, U.; Shutler, J.D.; Holding, T.; Ashton, I.G.C.; Landschutzer, P.; Woolf, D.K.; Goddijn-Murphy, L. Revised estimates of ocean-atmosphere CO_2 flux are consistent with ocean carbon inventory. *Nat. Commun.* **2020**, *11*, 4422. [CrossRef] [PubMed]
2. Feely, R.A.; Sabine, C.L.; Lee, K.; Berelson, W.; Kleypas, J.; Fabry, V.J.; Millero, F.J. Impact of anthropogenic CO_2 on the $CaCO_3$ system in the oceans. *Science* **2004**, *305*, 362–366. [CrossRef] [PubMed]
3. Sabine, C.L.; Feely, R.A.; Gruber, N.; Key, R.M.; Lee, K.; Bullister, J.L.; Wanninkhof, R.; Wong, C.S.; Wallace, D.W.; Tilbrook, B.; et al. The oceanic sink for anthropogenic CO_2. *Science* **2004**, *305*, 367–371. [CrossRef]
4. Orr, J.C.; Fabry, V.J.; Aumont, O.; Bopp, L.; Doney, S.C.; Feely, R.A.; Gnanadesikan, A.; Gruber, N.; Ishida, A.; Joos, F.; et al. Anthropogenic ocean acidification over the twenty-first century and its impact on calcifying organisms. *Nature* **2005**, *437*, 681–686. [CrossRef]
5. Hoegh-Guldberg, O.; Bruno, J.F. The impact of climate change on the world's marine ecosystems. *Science* **2010**, *328*, 1523–1528. [CrossRef] [PubMed]
6. Dupont, S.; Ortega-Martinez, O.; Thorndyke, M. Impact of near-future ocean acidification on echinoderms. *Ecotoxicology* **2010**, *19*, 449–462. [CrossRef] [PubMed]
7. Rodolfo-Metalpa, R.; Lombardi, C.; Cocito, S.; Hall-Spencer, J.M.; Gambi, M.C. Effects of ocean acidification and high temperatures on the bryozoan *Myriapora truncata* at natural CO_2 vents. *Mar. Ecol.* **2010**, *31*, 447–456. [CrossRef]
8. Matranga, V.; Bonaventura, R.; Costa, C.; Karakostis, K.; Pinsino, A.; Russo, R.; Zito, F. Echinoderms as Blueprints for Biocalcification: Regulation of Skeletogenic Genes and Matrices. In *Molecular Biomineralization: Aquatic Organisms Forming Extraordinary Materials*; Müller, W.E.G., Ed.; Springer: Berlin/Heidelberg, Germany, 2011; pp. 225–248.
9. Mann, S. *Mineralization in Biological Systems*; Springer: Berlin/Heidelberg, Germany, 1983; pp. 125–174.
10. Killian, C.E.; Wilt, F.H. Molecular aspects of biomineralization of the echinoderm endoskeleton. *Chem. Rev.* **2008**, *108*, 4463–4474. [CrossRef]
11. Gorzelak, P. *Functional Micromorphology of the Echinoderm Skeleton*; Cambridge University Press: Cambridge, UK, 2021; p. 44.
12. Feng, Q. Principles of Calcium-Based Biomineralization. In *Molecular Biomineralization: Aquatic Organisms Forming Extraordinary Materials*; Müller, W.E.G., Ed.; Springer: Berlin/Heidelberg, Germany, 2011; pp. 141–197.
13. Gilbert, P.U.P.A.; Wilt, F.H. Molecular Aspects of Biomineralization of the Echinoderm Endoskeleton. In *Molecular Biomineralization: Aquatic Organisms Forming Extraordinary Materials*; Müller, W.E.G., Ed.; Springer: Berlin/Heidelberg, Germany, 2011; pp. 199–223.
14. Ries, J.B.; Ghazaleh, M.N.; Connolly, B.; Westfield, I.; Castillo, K.D. Impacts of seawater saturation state (ΩA = 0.4–4.6) and temperature (10, 25 °C) on the dissolution kinetics of whole-shell biogenic carbonates. *Geochim. Cosmochim. Acta* **2016**, *192*, 318–337. [CrossRef]
15. Kawahata, H.; Fujita, K.; Iguchi, A.; Inoue, M.; Iwasaki, S.; Kuroyanagi, A.; Maeda, A.; Manaka, T.; Moriya, K.; Takagi, H.; et al. Perspective on the response of marine calcifiers to global warming and ocean acidification—Behavior of corals and foraminifera in a high CO_2 world "hot house". *Prog. Earth Planet Sci.* **2019**, *6*, 5. [CrossRef]
16. Morse, J.W.; Andersson, A.J.; Mackenzie, F.T. Initial responses of carbonate-rich shelf sediments to rising atmospheric pCO_2 and "ocean acidification": Role of high Mg-calcites. *Geochim. Cosmochim. Acta* **2006**, *70*, 5814–5830. [CrossRef]
17. Watson, S.A.; Peck, L.S.; Tyler, P.A.; Southgate, P.C.; Tan, K.S.; Day, R.W.; Morley, S.A. Marine invertebrate skeleton size varies with latitude, temperature and carbonate saturation: Implications for global change and ocean acidification. *Glob. Chang. Biol.* **2012**, *18*, 3026–3038. [CrossRef] [PubMed]
18. Anand, M.; Rangesh, K.; Maruthupandy, M.; Jayanthi, G.; Rajeswari, B.; Priya, R.J. Effect of CO_2 driven ocean acidification on calcification, physiology and ovarian cells of tropical sea urchin *Salmacis virgulata*—A microcosm approach. *Heliyon* **2021**, *7*, e05970. [CrossRef] [PubMed]
19. Duquette, A.; McClintock, J.B.; Amsler, C.D.; Perez-Huerta, A.; Milazzo, M.; Hall-Spencer, J.M. Effects of ocean acidification on the shells of four Mediterranean gastropod species near a CO_2 seep. *Mar. Pollut. Bull.* **2017**, *124*, 917–928. [CrossRef]
20. Ries, J.B.; Cohen, A.L.; McCorkle, D.C. Marine calcifiers exhibit mixed responses to CO_2-induced ocean acidification. *Geology* **2009**, *37*, 1131–1134. [CrossRef]
21. Pörtner, H. Ecosystem effects of ocean acidification in times of ocean warming: A physiologist's view. *Mar. Ecol. Prog. Ser.* **2008**, *373*, 203–217. [CrossRef]
22. Weiner, S.; Dove, P.M. An overview of biomineralization processes and the problem of the vital effect. *Rev. Mineral. Geochem.* **2003**, *54*, 1–29. [CrossRef]
23. Weber, J.N. Temperature dependence of magnesium in echinoid and asteroid skeletal calcite: A reinterpretation of its significance. *J. Geol.* **1973**, *81*, 543–556. [CrossRef]
24. Chave, K.E. Aspects of the biogeochemistry of magnesium 1. Calcareous marine organisms. *J. Geol.* **1954**, *62*, 266–283. [CrossRef]
25. Hermans, J.; Borremans, C.; Willenz, P.; André, L.; Dubois, P. Temperature, salinity and growth rate dependences of Mg/Ca and Sr/Ca ratios of the skeleton of the sea urchin *Paracentrotus lividus* (Lamarck): An experimental approach. *Mar. Biol.* **2010**, *157*, 1293–1300. [CrossRef]

26. Duquette, A.; Vohra, Y.K.; McClintock, J.B.; Angus, R.A. Near-future temperature reduces Mg/Ca ratios in the major skeletal components of the common subtropical sea urchin *Lytechinus variegatus*. *J. Exp. Mar. Biol. Ecol.* **2018**, *509*, 1–7. [CrossRef]
27. Lowenstam, H.A.; Weiner, S. *On Biomineralization*; Oxford University Press: New York, NY, USA, 1989.
28. Gazeau, F.; Parker, L.M.; Comeau, S.; Gattuso, J.-P.; O'Connor, W.A.; Martin, S.; Pörtner, H.-O.; Ross, P.M. Impacts of ocean acidification on marine shelled molluscs. *Mar. Biol.* **2013**, *160*, 2207–2245. [CrossRef]
29. Olson, I.C.; Kozdon, R.; Valley, J.W.; Gilbert, P.U. Mollusk shell nacre ultrastructure correlates with environmental temperature and pressure. *J. Am. Chem. Soc.* **2012**, *134*, 7351–7358. [CrossRef] [PubMed]
30. Reynaud, S.; Ferrier-Pagès, C.; Meibom, A.; Mostefaoui, S.; Mortlock, R.; Fairbanks, R.; Allemand, D. Light and temperature effects on Sr/Ca and Mg/Ca ratios in the scleractinian coral *Acropora* sp. *Geochim. Cosmochim. Acta* **2007**, *71*, 354–362. [CrossRef]
31. Toyofuku, T.; Kitazato, H.; Kawahata, H.; Tsuchiya, M.; Nohara, M. Evaluation of Mg/Ca thermometry in foraminifera: Comparison of experimental results and measurements in nature. *Paleoceanography* **2000**, *15*, 456–464. [CrossRef]
32. Dissard, D.; Nehrke, G.; Reichart, G.J.; Bijma, J. The impact of salinity on the Mg/Ca and Sr/Ca ratio in the benthic foraminifera *Ammonia tepida*: Results from culture experiments. *Geochim. Cosmochim. Acta* **2010**, *74*, 928–940. [CrossRef]
33. Bell, T.; Nishida, K.; Ishikawa, K.; Suzuki, A.; Nakamura, T.; Sakai, K.; Ohno, Y.; Iguchi, A.; Yokoyama, Y. Temperature-controlled culture experiments with primary polyps of coral *Acropora digitifera*: Calcification rate variations and skeletal Sr/Ca, Mg/Ca, and Na/Ca ratios. *Palaeogeogr. Palaeoclimatol. Palaeoecol.* **2017**, *484*, 129–135. [CrossRef]
34. Kısakürek, B.; Eisenhauer, A.; Böhm, F.; Garbe-Schönberg, D.; Erez, J. Controls on shell Mg/Ca and Sr/Ca in cultured planktonic foraminiferan, *Globigerinoides ruber* (white). *Earth Planet Sci. Lett.* **2008**, *273*, 260–269. [CrossRef]
35. Rathburn, A.E.; De Deckker, P. Magnesium and strontium compositions of recent benthic foraminifera from the Coral Sea, Australia and Prydz Bay, Antarctica. *Mar. Micropaleontol.* **1997**, *32*, 231–248. [CrossRef]
36. Dodd, J.R. Magnesium and strontium in calcareous skeletons: A review. *J. Paleontol.* **1967**, *41*, 1313–1329.
37. Kolbuk, D.; Di Giglio, S.; M'Zoudi, S.; Dubois, P.; Stolarski, J.; Gorzelak, P. Effects of seawater Mg^{2+}/Ca^{2+} ratio and diet on the biomineralization and growth of sea urchins and the relevance of fossil echinoderms to paleoenvironmental reconstructions. *Geobiology* **2020**, *18*, 710–724. [CrossRef] [PubMed]
38. LaVigne, M.; Hill, T.M.; Sanford, E.; Gaylord, B.; Russell, A.D.; Lenz, E.A.; Hosfelt, J.D.; Young, M.K. The elemental composition of purple sea urchin (*Strongylocentrotus purpuratus*) calcite and potential effects of pCO_2 during early life stages. *Biogeosciences* **2013**, *10*, 3465–3477. [CrossRef]
39. Todgham, A.E.; Stillman, J.H. Physiological responses to shifts in multiple environmental stressors: Relevance in a changing world. *Integr. Comp. Biol.* **2013**, *53*, 539–544. [CrossRef] [PubMed]
40. Harvey, B.P.; Gwynn-Jones, D.; Moore, P.J. Meta-analysis reveals complex marine biological responses to the interactive effects of ocean acidification and warming. *Ecol. Evol.* **2013**, *3*, 1016–1030. [CrossRef] [PubMed]
41. Kroeker, K.J.; Kordas, R.L.; Crim, R.; Hendriks, I.E.; Ramajo, L.; Singh, G.S.; Duarte, C.M.; Gattuso, J.-P. Impacts of ocean acidification on marine organisms: Quantifying sensitivities and interaction with warming. *Glob. Chang. Biol.* **2013**, *19*, 1884–1896. [CrossRef] [PubMed]
42. Brahmi, C.; Chapron, L.; Le Moullac, G.; Soyez, C.; Beliaeff, B.; Lazareth, C.E.; Gaertner-Mazouni, N.; Vidal-Dupiol, J. Effects of elevated temperature and pCO_2 on the respiration, biomineralization and photophysiology of the giant clam *Tridacna maxima*. *Conserv. Physiol.* **2021**, *9*, coab041. [CrossRef]
43. Knights, A.M.; Norton, M.J.; Lemasson, A.J.; Stephen, N. Ocean acidification mitigates the negative effects of increased sea temperatures on the biomineralization and crystalline ultrastructure of *Mytilus*. *Front. Mar. Sci.* **2020**, *7*. [CrossRef]
44. Li, S.; Liu, C.; Huang, J.; Liu, Y.; Zheng, G.; Xie, L.; Zhang, R. Interactive effects of seawater acidification and elevated temperature on biomineralization and amino acid metabolism in the mussel *Mytilus edulis*. *J. Exp. Biol.* **2015**, *218*, 3623–3631. [CrossRef]
45. Byrne, M.; Smith, A.M.; West, S.; Collard, M.; Dubois, P.; Graba-landry, A.; Dworjanyn, S.A. Warming influences Mg^{2+} content, while warming and acidification influence calcification and test strength of a sea urchin. *Environ. Sci. Technol.* **2014**, *48*, 12620–12627. [CrossRef]
46. Byrne, M. Impact of ocean warming and ocean acidification on marine invertebrate life history stages: Vulnerabilities and potential for persistence in a changing ocean. In *Oceanography and Marine Biology: An Annual Review*; Gibson, R.N., Atkinson, R.J.A., Gordon, J.D.M., Smith, I.P., Hughes, D.J., Eds.; Taylor & Francis: Boca Raton, FL, USA, 2011; pp. 1–42.
47. Politi, Y.; Arad, T.; Klein, E.; Weiner, S.; Addadi, L. Sea urchin spine calcite forms via a transient amorphous calcium carbonate phase. *Science* **2004**, *306*, 1161–1164. [CrossRef]
48. Kokorin, A.I.; Mirantsev, G.V.; Rozhnov, S.V. General features of echinoderm skeleton formation. *Paleontol. J.* **2015**, *48*, 1532–1539. [CrossRef]
49. Borremans, C.; Hermans, J.; Baillon, S.; André, L.; Dubois, P. Salinity effects on the Mg/Ca and Sr/Ca in starfish skeletons and the echinoderm relevance for paleoenvironmental reconstructions. *Geology* **2009**, *37*, 351–354. [CrossRef]
50. Dubois, P. The skeleton of postmetamorphic echinoderms in a changing world. *Biol. Bull.* **2014**, *226*, 223–236. [CrossRef] [PubMed]
51. Weiner, S. Organic matrixlike macromolecules associated with the mineral phase of sea urchin skeletal plates and teeth. *J. Exp. Zool.* **1985**, *234*, 7–15. [CrossRef] [PubMed]
52. Killian, C.E.; Wilt, F.H. Characterization of the proteins comprising the integral matrix of *Strongylocentrotus purpuratus* embryonic spicules. *J. Biol. Chem.* **1996**, *271*, 9150–9159. [CrossRef]

53. Addadi, L.; Weiner, S. Control and design principles in biological mineralization. *Angew. Chem. Int. Ed.* **1992**, *31*, 153–169. [CrossRef]
54. Hermans, J.; Andre, L.; Navez, J.; Pernet, P.; Dubois, P. Relative influences of solution composition and presence of intracrystalline proteins on magnesium incorporation in calcium carbonate minerals: Insight into vital effects. *J. Geophys. Res.* **2011**, *116*, G01001. [CrossRef]
55. Ries, J.B. A physicochemical framework for interpreting the biological calcification response to CO_2-induced ocean acidification. *Geochim. Cosmochim. Acta* **2011**, *75*, 4053–4064. [CrossRef]
56. Mackenzie, F.T.; Bischoff, W.D.; Bishop, F.C.; Loijens, M.; Schoonmaker, J.; Wollast, R. Magnesian Calcites: Low Temperature Occurrence, Solubility and Solid Solution Behavior. In *Carbonates: Mineralogy and Chemistry, Reviews in Mineralogy*; Richard, J.R., Ed.; Mineralogical Society of America: Stony Brook, NY, USA, 1983; Volume 11, pp. 97–144.
57. Knoll, A.H. Biomineralization and evolutionary history. *Rev. Mineral. Geochem.* **2003**, *54*, 329–356. [CrossRef]
58. Byrne, M.; Fitzer, S. The impact of environmental acidification on the microstructure and mechanical integrity of marine invertebrate skeletons. *Conserv. Physiol.* **2019**, *7*, coz062. [CrossRef]
59. Andersson, A.J.; Mackenzie, F.T.; Bates, N.R. Life on the margin: Implications of ocean acidification on Mg-calcite, high latitude and cold-water marine calcifiers. *Mar. Ecol. Prog. Ser.* **2008**, *373*, 265–273. [CrossRef]
60. O'Loughlin, P.M.; Rowe, F.W.E. A systematic revision of the asterinid genus *Aquilonastra* O'Loughlin, 2004 (Echinodermata: Asteroidea). *Mem. Mus. Vic.* **2006**, *63*, 257–287. [CrossRef]
61. Menge, B.A.; Sanford, E. Ecological role of sea stars from populations to meta ecosystems. In *Starfish: Biology and Ecology of the Asteroidea*; Lawrence, J.M., Ed.; The Johns Hopkins University Press: Baltimore, MD, USA, 2013; pp. 67–80.
62. Balogh, R.; Byrne, M. Developing in a warming intertidal, negative carry over effects of heatwave conditions in development to the pentameral starfish in *Parvulastra exigua*. *Mar. Environ. Res.* **2020**, *162*, 105083. [CrossRef] [PubMed]
63. Nguyen, H.D.; Byrne, M. Early benthic juvenile *Parvulastra exigua* (Asteroidea) are tolerant to extreme acidification and warming in its intertidal habitat. *J. Exp. Mar. Biol. Ecol.* **2014**, *453*, 36–42. [CrossRef]
64. McElroy, D.J.; Nguyen, H.D.; Byrne, M. Respiratory response of the intertidal seastar *Parvulastra exigua* to contemporary and near-future pulses of warming and hypercapnia. *J. Exp. Mar. Biol. Ecol.* **2012**, *416–417*, 1–7. [CrossRef]
65. Khalil, M.; Doo, S.S.; Stuhr, M.; Westphal, H. Leibniz Centre for Tropical Marine Research (ZMT), Bremen, Germany. 2022, Unpublished Work.
66. Dickson, A.G.; Sabine, C.L.; Christian, J.R. (Eds.) *Guide to Best Practices for Ocean CO_2 Measurements*; PICES Special Publication 3; North Pacific Marine Science Organization: British Columbia, Canada, 2007; Volume 3, p. 191.
67. Bray, L.; Pancucci-Papadopoulou, M.A.; Hall-Spencer, J.M. Sea urchin response to rising pCO_2 shows ocean acidification may fundamentally alter the chemistry of marine skeletons. *Mediterr. Mar. Sci.* **2014**, *15*, 510–519. [CrossRef]
68. Ehrlich, H.; Elkin, Y.N.; Artoukov, A.A.; Stonik, V.A.; Safronov, P.P.; Bazhenov, V.V.; Kurek, D.V.; Varlamov, V.P.; Born, R.; Meissner, H.; et al. Simple method for preparation of nanostructurally organized spines of sand dollar *Scaphechinus mirabilis* (Agassiz, 1863). *Mar. Biotechnol.* **2011**, *13*, 402–410. [CrossRef]
69. Russell, A.D.; Hönisch, B.; Spero, H.J.; Lea, D.W. Effects of seawater carbonate ion concentration and temperature on shell U, Mg, and Sr in cultured planktonic foraminifera. *Geochim. Cosmochim. Acta* **2004**, *68*, 4347–4361. [CrossRef]
70. R Core Team. *R: A Language and Environment for Statistical Computing*; R Foundation for Statistical Computing: Vienna, Austria, 2022.
71. Shapiro, S.S.; Wilk, M.B. An analysis of variance test for normality (complete samples). *Biometrika* **1965**, *52*, 591–611. [CrossRef]
72. Levene, H. Robust tests for equality of variances. In *Contributions to Probability and Statistics*; Olkin, I., Ed.; Stanford University Press: Palo Alto, CA, USA, 1960; pp. 278–292.
73. De Mendiburu, F. Agricolae: Statistical Procedures for Agricultural Research. R Package Version 1.3-5. 2021. Available online: https://CRAN.R-project.org/package=agricolae (accessed on 19 January 2022).
74. Weber, J.N. The incorporation of magnesium into the skeletal calcites of echinoderms. *Am. J. Sci.* **1969**, *267*, 537–566. [CrossRef]
75. Rosenthal, Y.; Boyle, E.A.; Slowey, N. Temperature control on the incorporation of magnesium, strontium, fluorine, and cadmium into benthic foraminiferal shells from Little Bahama Bank: Prospects for thermocline paleoceanography. *Geochim. Cosmochim. Acta* **1997**, *61*, 3633–3643. [CrossRef]
76. Purgstaller, B.; Mavromatis, V.; Goetschl, K.E.; Steindl, F.R.; Dietzel, M. Effect of temperature on the transformation of amorphous calcium magnesium carbonate with near-dolomite stoichiometry into high Mg-calcite. *CrystEngComm* **2021**, *23*, 1969–1981. [CrossRef]
77. Radha, A.V.; Fernandez-Martinez, A.; Hu, Y.; Jun, Y.-S.; Waychunas, G.A.; Navrotsky, A. Energetic and structural studies of amorphous $Ca_{1-x}Mg_xCO_3 \cdot nH_2O$ ($0 \leq x \leq 1$). *Geochim. Cosmochim. Acta* **2012**, *90*, 83–95. [CrossRef]
78. Di Lorenzo, F.; Rodríguez-Galán, R.M.; Prieto, M. Kinetics of the solvent-mediated transformation of hydromagnesite into magnesite at different temperatures. *Mineral. Mag.* **2018**, *78*, 1363–1372. [CrossRef]
79. Stumpp, M.; Hu, M.Y. pH Regulation and Excretion in Echinoderms. In *Acid-Base Balance and Nitrogen Excretion in Invertebrates*; Weihrauch, D., O'Donnell, M., Eds.; Springer: Berlin/Heidelberg, Germany, 2017; pp. 261–273.
80. Miles, H.; Widdicombe, S.; Spicer, J.I.; Hall-Spencer, J. Effects of anthropogenic seawater acidification on acid-base balance in the sea urchin *Psammechinus miliaris*. *Mar. Pollut. Bull.* **2007**, *54*, 89–96. [CrossRef] [PubMed]

81. Asnaghi, V.; Mangialajo, L.; Gattuso, J.-P.; Francour, P.; Privitera, D.; Chiantore, M. Effects of ocean acidification and diet on thickness and carbonate elemental composition of the test of juvenile sea urchins. *Mar. Environ. Res.* **2014**, *93*, 78–84. [CrossRef]
82. Lorens, R.B. Sr, Cd, Mn and Co distribution coefficients in calcite as a function of calcite precipitation rate. *Geochim. Cosmochim. Acta* **1981**, *45*, 553–561. [CrossRef]
83. Lea, D.W.; Mashiotta, T.A.; Spero, H.J. Controls on magnesium and strontium uptake in planktonic foraminifera determined by live culturing. *Geochim. Cosmochim. Acta* **1999**, *63*, 2369–2379. [CrossRef]
84. Stoll, H.M.; Rosenthal, Y.; Falkowski, P. Climate proxies from Sr/Ca of coccolith calcite: Calibrations from continuous culture of *Emiliania huxleyi*. *Geochim. Cosmochim. Acta* **2002**, *66*, 927–936. [CrossRef]
85. Shannon, R.D. Revised effective ionic radii and systematic studies of interatomic distances in halides and chalcogenides. *Acta Cryst.* **1976**, *32*, 751–767. [CrossRef]
86. Tesoriero, A.J.; Pankow, J.F. Solid solution partitioning of Sr^{2+}, Ba^{2+}, and Cd^{2+} to calcite. *Geochim. Cosmochim. Acta* **1996**, *60*, 1053–1063. [CrossRef]
87. Mucci, A.; Morse, J.W. The incorporation of Mg^{2+} and Sr^{2+} into calcite overgrowths: Influences of growth rate and solution composition. *Geochim. Cosmochim. Acta* **1983**, *47*, 217–233. [CrossRef]
88. Zeebe, R.E.; Wolf-Gladrow, D. *CO_2 in Seawater: Equilibrium, Kinetics, Isotopes*; Elsevier: Amsterdam, The Netherlands, 2001.
89. Dubois, P.; Chen, C.P. Calcification in echinoderms. In *Echinoderm Studies*; Jangoux, M., Lawrence, J.M., Eds.; AA Balkema: Rotterdam, The Netherlands, 1989; Volume 3, pp. 109–178.
90. Melzner, F.; Mark, F.C.; Seibel, B.A.; Tomanek, L. Ocean acidification and coastal marine invertebrates: Tracking CO_2 effects from seawater to the cell. *Ann. Rev. Mar. Sci.* **2020**, *12*, 499–523. [CrossRef] [PubMed]
91. Feely, R.; Doney, S.; Cooley, S. Ocean acidification: Present conditions and future changes in a high-CO_2 world. *Oceanography* **2009**, *22*, 36–47. [CrossRef]
92. Challener, R.C.; McClintock, J.B.; Makowsky, R. Effects of reduced carbonate saturation state on early development in the common edible sea urchin *Lytechinus variegatus*: Implications for land-based aquaculture. *J. Appl. Aquac.* **2013**, *25*, 154–175. [CrossRef]
93. McClintock, J.B.; Amsler, M.O.; Angus, R.A.; Challener, R.C.; Schram, J.B.; Amsler, C.D.; Mah, C.L.; Cuce, J.; Baker, B.J. The Mg-Calcite composition of antarctic echinoderms: Important implications for predicting the impacts of ocean acidification. *J. Geol.* **2011**, *119*, 457–466. [CrossRef]
94. Johnson, R.; Harianto, J.; Thomson, M.; Byrne, M. The effects of long-term exposure to low pH on the skeletal microstructure of the sea urchin *Heliocidaris erythrogramma*. *J. Exp. Mar. Biol. Ecol.* **2020**, *523*, 151250. [CrossRef]
95. Hazan, Y.; Wangensteen, O.S.; Fine, M. Tough as a rock-boring urchin: Adult *Echinometra* sp. EE from the Red Sea show high resistance to ocean acidification over long-term exposures. *Mar. Biol.* **2014**, *161*, 2531–2545. [CrossRef]
96. Gibbin, E.M.; Massamba N'Siala, G.; Chakravarti, L.J.; Jarrold, M.D.; Calosi, P. The evolution of phenotypic plasticity under global change. *Sci. Rep.* **2017**, *7*, 17253. [CrossRef]
97. Stumpp, M.; Hu, M.Y.; Melzner, F.; Gutowska, M.A.; Dorey, N.; Himmerkus, N.; Holtmann, W.C.; Dupont, S.T.; Thorndyke, M.C.; Bleich, M. Acidified seawater impacts sea urchin larvae pH regulatory systems relevant for calcification. *Proc. Natl. Acad. Sci. USA* **2012**, *109*, 18192–18197. [CrossRef]
98. Wood, H.L.; Spicer, J.I.; Widdicombe, S. Ocean acidification may increase calcification rates, but at a cost. *Proc. Biol. Sci.* **2008**, *275*, 1767–1773. [CrossRef] [PubMed]
99. Pelletier, G.; Lewis, E.; Wallace, D. CO2SYS.xls: A calculator for the CO2 System in Seawater for Microsoft Excel/VBA. 2007. Available online: https://www.baltex-research.eu/baltic2009/downloads/Lectures/Bernd_Schneider/co2sys_ver14.xls (accessed on 19 January 2022).
100. Hansson, I. A new set of acidity constants for carbonic acid and boric acid in sea water. *Deep Sea Res. Oceanogr. Abstr.* **1973**, *20*, 461–478. [CrossRef]
101. Mehrbach, C.; Culberson, C.H.; Hawley, J.E.; Pytkowicx, R.M. Measurement of the apparent dissociation constants of carbonic acid in seawater at atmospheric pressure. *Limnol. Oceanogr.* **1973**, *18*, 897–907. [CrossRef]
102. Dickson, A.; Millero, F. A comparison of the equilibrium constants for the dissociation of carbonic acid in seawater media. *Deep Sea Res. Part A Oceanogr. Res. Pap.* **1987**, *34*, 1733–1743. [CrossRef]
103. Dickson, A.G. Standard potential of the reaction: AgCl(s) + 1/2H2(g) = Ag(s) + HCl(aq), and the standard acidity constant of the ion HSO4- in synthetic seawater from 273.15 to 318.15 K. *J. Chem. Thermodyn.* **1990**, *22*, 113–127. [CrossRef]
104. Dickson, A.; Riley, J. The estimation of acid dissociation constants in seawater media from potentiometric titrations with strong base. I. The ionic product of water—Kw. *Mar. Chem.* **1979**, *7*, 89–99. [CrossRef]
105. Uppström, L.R. The boron/chlorinity ratio of deep-sea water from the Pacific Ocean. *Deep Sea Res. Oceanogr. Abstr.* **1974**, *21*, 161–162. [CrossRef]
106. Mucci, A. The solubility of calcite and aragonite in seawater at various salinities, temperatures, and one atmosphere total pressure. *Am. J. Sci.* **1983**, *283*, 780–799. [CrossRef]

Journal of *Marine Science and Engineering*

MDPI

Article

Responses of Freshwater Calcifiers to Carbon-Dioxide-Induced Acidification

Aaron T. Ninokawa [1,2,*] and Justin Ries [3,4]

1 Bodega Marine Laboratory, University of California, Davis, Bodega Bay, CA 94923, USA
2 Friday Harbor Laboratories, University of Washington, Friday Harbor, WA 98250, USA
3 Department of Marine and Environmental Sciences, Northeastern University, Boston, NA 02115, USA
4 Tahoe Environmental Research Center, University of California, Davis, Incline Village, NV 89451, USA
* Correspondence: anino@uw.edu

Abstract: Increased anthropogenic carbon dioxide (CO_2) in the atmosphere can enter surface waters and depress pH. In marine systems, this phenomenon, termed ocean acidification (OA), can modify a variety of physiological, ecological, and chemical processes. Shell-forming organisms are particularly sensitive to this chemical shift, though responses vary amongst taxa. Although analogous chemical changes occur in freshwater systems via absorption of CO_2 into lakes, rivers, and streams, effects on freshwater calcifiers have received far less attention, despite the ecological importance of these organisms to freshwater systems. We exposed four common and widespread species of freshwater calcifiers to a range of pCO_2 conditions to determine how CO_2-induced reductions in freshwater pH impact calcium carbonate shell formation. We incubated the signal crayfish, *Pacifastacus leniusculus*, the Asian clam, *Corbicula fluminea*, the montane pea clam, *Pisidium* sp., and the eastern pearlshell mussel, *Margaritifera margaritifera*, under low pCO_2 conditions ($pCO_2 = 616 \pm 151$ µatm; pH = 7.91 ± 0.11), under moderately elevated pCO_2 conditions ($pCO_2 = 1026 \pm 239$ uatm; pH = 7.67 ± 0.10), and under extremely elevated pCO_2 conditions ($pCO_2 = 2380 \pm 693$ uatm; pH = 7.32 ± 0.12). Three of these species exhibited a negative linear response to increasing pCO_2 (decreasing pH), while the fourth, the pea clam, exhibited a parabolic response. Additional experiments revealed that feeding rates of the crayfish decreased under the highest pCO_2 treatment, potentially contributing to or driving the negative calcification response of the crayfish to elevated pCO_2 by depriving them of energy needed for biocalcification. These results highlight the potential for freshwater taxa to be deleteriously impacted by increased atmospheric pCO_2, the variable nature of these responses, and the need for further study of this process in freshwater systems.

Keywords: ocean acidification; freshwater acidification; freshwater calcifier; carbon dioxide; calcification; biomineralization; Lake Tahoe

Citation: Ninokawa, A.T.; Ries, J. Responses of Freshwater Calcifiers to Carbon-Dioxide-Induced Acidification. *J. Mar. Sci. Eng.* **2022**, *10*, 1068. https://doi.org/10.3390/jmse10081068

Academic Editors: Maria Gabriella Marin and Ryan J.K. Dunn

Received: 21 May 2022
Accepted: 27 July 2022
Published: 4 August 2022

1. Introduction

Increased partial pressure of atmospheric carbon dioxide (pCO_2) drives large scale alterations to environmental systems. In particular, this additional CO_2 can enter surface waters and perturb the aquatic carbonate system, lowering pH, carbonate ion concentration ($[CO_3^{2-}]$), and the saturation state of water with respect to calcium carbonate ($CaCO_3$). In marine systems, this process, termed ocean acidification, can impair tissue and shell growth and alter the behavior of many marine species, yielding ecological changes across a range of spatial, temporal, and trophic scales [1].

Calcifying organisms, those producing calcium carbonate shells or skeletons, are especially sensitive to CO_2-induced changes in carbonate system chemistry, although the drivers of this sensitivity can be complex [2–4]. In many taxa, calcium carbonate material is produced more slowly under increased pCO_2 [1,5], with shell and skeletal material produced under these conditions tending to be weaker [6,7]. Some species, however, grow faster under increased pCO_2 or can maintain or enhance the quality of their shell or skeletal material [5,8]. In many cases, the effects of pCO_2 operate indirectly, often via pathways related to pH, bicarbonate, or carbonate ions [3,9]. Indeed, the diversity of responses to increased pCO_2 complicates the process of generalizing how calcifying taxa will respond to CO_2-induced acidification in the future [4].

Although there have been many studies examining effects of CO_2-induced acidification on marine species, far fewer have considered responses of freshwater species. Freshwater typically has lower alkalinity, or chemical buffering capacity, which makes the $CaCO_3$ saturation state of freshwater less than that of seawater for equivalent pCO_2 conditions, while also rendering the carbonate chemistry of freshwater systems more sensitive to variations in atmospheric pCO_2 than that of marine systems [10,11]. Freshwater species, for example, will experience larger shifts in pH and lower absolute $CaCO_3$ saturation states than their marine analogs for a given increase in anthropogenic CO_2. The lower alkalinity of freshwater systems also makes the carbonate chemistry of these systems more vulnerable to diurnal and seasonal cycles in photosynthesis and respiration than marine systems [10]. It is, therefore, possible that freshwater calcifiers will be more sensitive than marine calcifiers to equivalent shifts in atmospheric pCO_2. Alternatively, the naturally higher variability in carbonate chemistry and lower absolute saturation state of freshwater systems could select for freshwater calcifying taxa that are relatively resilient to CO_2-induced perturbations to carbonate system chemistry.

Although past research has examined responses of freshwater calcifying species to elevated aqueous CO_2, most of this earlier work was conducted with the aim of understanding physiological acid–base dynamics within the organisms themselves, rather than understanding the impacts of CO_2-induced acidification on shell and skeletal formation. Thus, much of this prior work employed unrealistically high pCO_2 conditions that exceed projected end-of-century CO_2 partial pressures by tens of thousands of microatmospheres (μatm), resulting in unrealistic pH reductions in the order of 2 units [12]. Furthermore, among the modest number of studies where pCO_2 treatments reflected realistic future scenarios, prior work has focused on shifts in primary productivity by phytoplankton and changes in food quality [10] as driving calcifier performance, rather than on the direct effects of CO_2-induced changes in freshwater carbonate chemistry. Therefore, to determine the effect of increased atmospheric pCO_2 on calcifying invertebrates in the context of increasing anthropogenic pCO_2, we reared four species of freshwater calcifiers under three atmospheric pCO_2 conditions that bracket the range of conditions expected to occur over the next two centuries given the range of future emissions scenarios [13].

2. Materials and Methods

This project was conducted at the University of California, Davis, Tahoe Environmental Research Center (Tahoe City Field Station) and Lot #4 of Alpine Meadows, CA, USA between December 2020 and August 2021. The freshwater employed in the flow-through CO_2-induced acidification experiments was sourced from Burton Spring, a tributary of Lake Tahoe. The carbonate chemistry of the treatments was controlled by maintaining constant pH in three 800 L sumps with a pH-stat system (American Marine Inc. Pinpoint pH Controller) that reduced pH by dosing pure compressed CO_2 with a solenoid-valve controlled gas regulator (FZone Pro Series CO_2 Regulator), or increased pH by dosing ambient air through an air pump (Simply Deluxe Electromagnetic Air Pump). The compressed CO_2 and compressed air were sparged into the sumps and treatments tanks with flexible, microporous bubbling tubes that were designed to expedite equilibration between aqueous and gas phases. Spring water (filtered to 10 μm) entered the sumps at a rate of 1 L min^{-1} to minimize accumulation of metabolic byproducts and prevent depletion of alkalinity and calcium ions through the calcification process. The sumps and experimental tanks were covered with plastic sheeting and plastic lids, respectively, to prevent room air from equilibrating with the water in the sumps and the experimental treatments. The water within each of the three sump systems was recirculated through 12–40 L tanks (3 sumps × 12 tanks = 36 tanks total), with the recirculating water passing through an activated carbon filter (changed biweekly) before entering the tanks and through 10 L of loose activated carbon media (changed monthly) before re-entering the sump (Figure 1). Temperature varied naturally with the seasons but was maintained below 22 °C with chillers and the addition of frozen spring water.

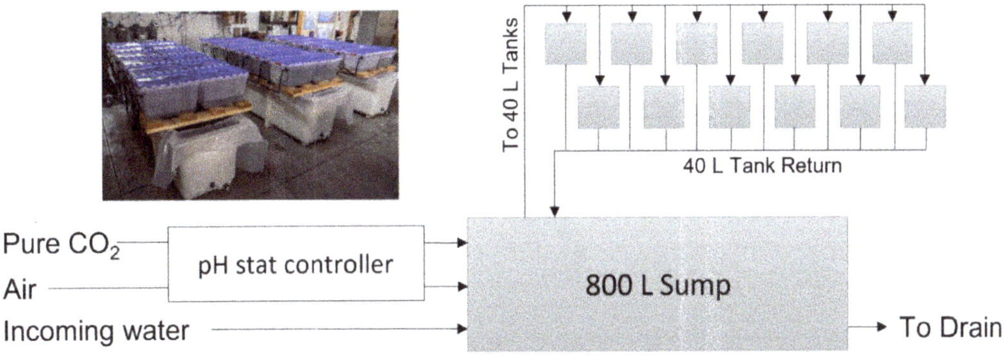

Figure 1. Schematic diagram of one of the three identical pCO_2 control systems used in the experiment. Fresh spring water flowed continuously through the main sump and recirculated amongst the experimental tanks. A pH stat controller added either pure CO_2 or ambient air as required to maintain water pH within the target range. The image in the top left shows arrangement of all three pCO_2 treatment systems in the laboratory.

Water chemistry was measured three times per week with a Thermo Scientific Orion Star A329 Multiparameter Meter. Oxygen (Orion RDO Optical Dissolved Oxygen Probe) and pH sensors (Atlas Scientific Spear-tip pH) were calibrated daily in air-equilibrated water and with NIST-traceable NBS buffers (pH 7 and 10), respectively. The conductivity probe (Duraprobe) was calibrated every other week with Oakton 84 and 1413 $\mu S\ cm^{-1}$ standards. Discrete 200 mL water samples were collected from the three sumps during each of these sampling events and frozen until they could be analyzed for alkalinity with a Metrohm 855 Robotic Titrosampler.

In addition to these higher frequency measurements, discrete 500 mL water samples were collected weekly from each of the 36 tanks and analyzed at the Tahoe Environmental Research Center (Incline Village Field Station) for dissolved inorganic carbon (DIC) on a Lachat IL 500 TOC instrument. The 'seacarb' package in *R* was used to calculate total alkalinity (TA) from these lower frequency DIC and pH measurements using constants from Waters et al. [14]. This information was used to define the linear relationship between specific conductance and TA for water employed in this experiment (Figure 2). We then used this relationship to estimate TA from the higher frequency measurement of water-specific conductance (described above). The calculated TA was 96 ± 0.3% (mean ± se, n = 149) of the measured alkalinity described above. Given this good agreement between calculated and measured alkalinity, we then used these higher frequency measurements of pH and conductivity-estimated alkalinity, along with measured temperature and total dissolved solids, to calculate carbonate system parameters for each tank throughout the duration of the experiment. The saturation state of calcite ($\Omega_{calcite}$) was calculated assuming that the calcium concentration was half the alkalinity as calcium concentrations in the Lake Tahoe area tend to be between 0.5 and 1 times the alkalinity [15]. Small deviations in Ca^{2+} concentration from this relationship will not materially impact the calculated saturation states or the interpretation of the results.

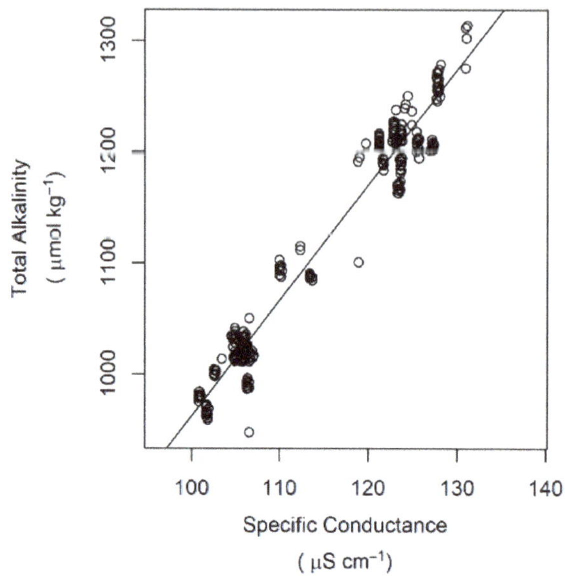

Figure 2. Relationship between specific conductance and total alkalinity for water samples obtained from this experiment. The linearity of this relationship allowed the calculation of total alkalinity (for carbonate system calculations) from higher frequency measurements of specific conductance.

The species investigated in this experiment were collected from the field and acclimated to laboratory conditions for at least 5 days prior to starting the experiments. Signal crayfish, *Pacifastacus leniusculus*, were collected from Pomin Park in Tahoe City, CA, USA, in late April 2021. Two cohorts of Asian clams, *Corbicula fluminea*, were collected from Marla Bay and Lakeside near Stateline, NV, on 15 April 2021 and 24 May 2021.

Two cohorts of pea clams, *Pisidium* sp., were collected from a spring feeding into the Truckee River near Polaris, CA, on 13 May 2021 and 19 June 2021. Pearlshell mussels, *Margaritifera margaritifera*, were collected from a lake near Jupiter, Florida, in early May 2021. Bivalves were held in three of the 40 L tanks per CO_2 treatment while crayfish were held in a different set of three tanks (9 tanks total per species). Remaining tanks (6 per pCO_2 treatment) were used for respiration and feeding trials described below. Bivalves were fed a commercially available Shellfish Diet twice daily at a concentration of 5 mL per 40 L of tank water. Crayfish were fed 1 × 1 cm dehydrated algae sheets and raw shrimp ad libitum every other day, with uneaten food removed at the time of feeding. All tanks were cleaned of accumulated solid waste three or four times per week.

Organisms remained within the tanks for the duration of the experiment and net calcification rates were calculated as the fractional change in shell mass determined from buoyant weights at the beginning and end of the experiment, normalized to a 30-day interval. Shell mass was calculated from buoyant weight measurements, where the density of the water was determined daily by the air and buoyant weights of glass bead standards (measured daily) with densities of either 2.55 or 2.23 g cm^{-3} [16]. Daily buoyant and dry weights of four half shells of *Corbicula fluminea* and two half shells of *Margaritifera margaritifera* revealed that these species had shell densities of 2.81 and 2.71 g cm^{-3}, respectively and these values were used for calculating shell mass from buoyant weight [16]. A shell density of 2.71 g cm^{-3} was used for calculating the crayfish shell mass from their buoyant weight. Because crayfish transport calcium carbonate between their external shell and internal gastrolith during the molting process [17], crayfish that, at the conclusion of the incubation, had recently molted and had not yet remineralized their shell (i.e., the shell was soft) were excluded from further analysis. Due to their small size, net calcification rate of pea clams was calculated as the fractional change in wet mass between the beginning and end of the experiment normalized to a 30-day interval, measured with a Sartorius 2120T microbalance after removing excess water from the shell with lint-free wipes and allowing the clams to air dry for precisely 15 min prior to weighing. Net calcification rates were analyzed with a maximum likelihood routine fitting various functional models identified previously [5], including linear, parabolic, and threshold (exponential) relationships with average pCO_2 during the incubation period using individuals as replicates. The statistically significant model with the lowest AIC was selected as the best one to describe the relationship between pCO_2 and net calcification rate.

At the conclusion of the growth experiments, additional experiments were conducted to test whether CO_2-induced changes in carbonate chemistry impaired crayfish feeding. A new batch of live crayfish prey (pea clams) was collected and acclimated to the pCO_2 treatments for three days. Five pea clams of comparable size were placed in each of six tanks per pCO_2 treatment with a single crayfish per tank. The number of pea clams consumed after each 25 min trial was recorded. The respiration rates of the crayfish were also obtained to determine whether pCO_2-induced changes in water chemistry altered metabolic rates and, therefore, feeding requirements. Before the feeding trials, we placed each crayfish in a sealed incubation vessel and measured the change in oxygen concentration during the incubation time with the Orion RDO dissolved oxygen probe. Respiration and feeding trials both took place at night in the dark, as this is when the crayfish are most active.

3. Results

The freshwater calcifying organisms in this study were exposed to pCO_2 conditions approximately double and quadruple those of average present-day conditions. Mean daily $pCO_2 \pm$ SD (calculated from pH, total alkalinity, TDS, and temperature) across all tanks in the control, intermediate, and high treatments were 616 ± 151 µatm, 1026 ± 239, and 2380 ± 693 (Figure 3d, Table 1, n = 35 sampling days), respectively, corresponding to mean daily pH $\pm$ SD of 7.91 ± 0.11, 7.67 ± 0.10, and 7.32 ± 0.12 (Figure 3b, Table 1, n = 35 sampling days). Daily mean alkalinity decreased throughout the course of the experiment from 1272 to 927 µmol kg^{-1} (Figure 3c), probably due to increased contribution of snowmelt relative to groundwater through the spring–summer transition. This trend was accompanied by an increase in daily mean temperature from 12.2 to 21.4 °C (Figure 3a). The relationship between specific conductance (SC, µS cm^{-1}) and total alkalinity (TA, µmol kg^{-1}) was $TA = -75.46 + 10.38 * SC$ ($R^2 = 0.96$, $F_{1314} = 6937$, $p < 0.001$).

Table 1. Average treatment conditions during the experiment. Measurements are the mean (standard deviation) of 12 tanks per treatment over 35 sampling days.

Parameter	Units	Low pCO_2	Intermediate pCO_2	High pCO_2
Temperature	°C	17.9 (2.6)	17.9 (2.5)	17.6 (2.5)
Specific conductance	µS cm^{-1}	113 (11)	111 (10)	113 (11)
Dissolved oxygen	µmol L^{-1}	232 (13)	230 (13)	227 (9)
pH (NBS scale)		7.91 (0.11)	7.67 (0.10)	7.32 (0.12)
Total alkalinity	µmol kg^{-1}	1098 (116)	1078 (102)	1093 (110)
pCO_2	µatm	616 (151)	1026 (239)	2380 (693)
$[HCO_3^-]$	µmol kg^{-1}	1088 (114)	1072 (102)	1090 (110)
$[CO_3^{2-}]$	µmol kg^{-1}	4.7 (1.4)	2.7 (0.7)	1.2 (0.4)
$\Omega_{aragonite}$		0.50 (0.16)	0.28 (0.08)	0.13 (0.04)

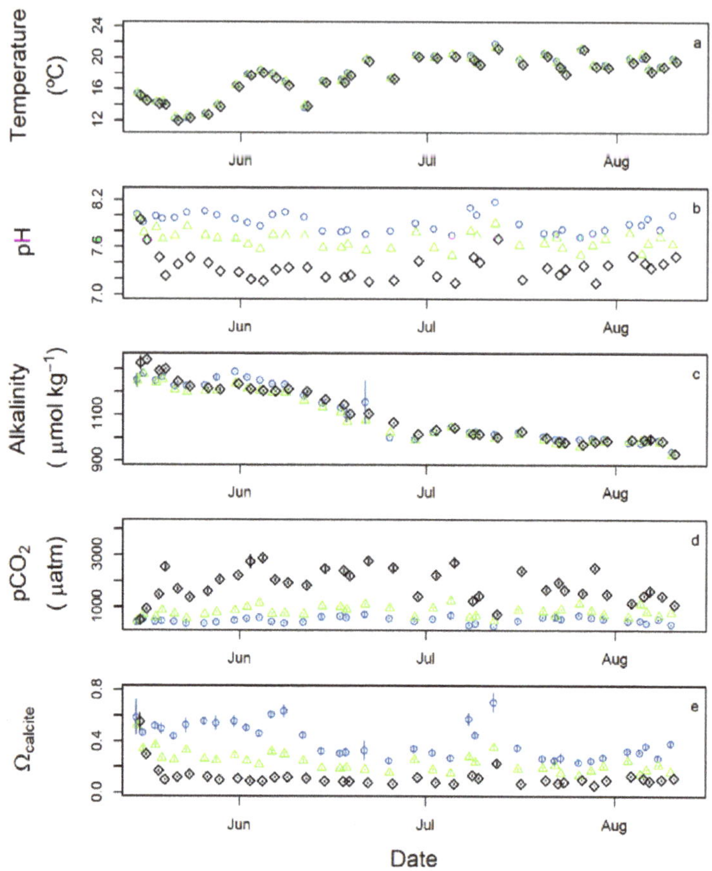

Figure 3. Average daily conditions within the experimental replicate tanks throughout the duration of the experiment. Blue circles indicate the low pCO_2 treatment, green triangles indicate the intermediate pCO_2 treatment, and black diamonds indicate the high pCO_2 treatment. Vertical bars show the standard deviation of all 12 tanks at each pCO_2 treatment.

Net calcification rates of the four species as a function of pCO_2 followed two general response types (Tables 2 and 3). Net calcification rates of *Pacifastacus leniusculus* (Figure 4a), *Corbicula fluminea* (Figure 4b), and *Margaritifera margaritifera* (Figure 4c) exhibited a linear decline in net calcification with increasing pCO_2 (decreasing pH). Net calcification rate of *Pisidium* sp. exhibited a parabolic relationship with pCO_2, in which maximum rates of net calcification were observed in the intermediate pCO_2 treatment (Figure 4d).

Table 2. Net calcification rates (fractional change in buoyant or wet weight normalized to a 30 day growth interval) in each treatment for each species during the experiment. Values reported are mean (standard error), sample size.

Species	Low pCO_2	Intermediate pCO_2	High pCO_2
P. leniusculus	−0.007 (0.003), 10	−0.009 (0.002), 15	−0.019 (0.003), 8
C. fluminea	−0.018 (0.004), 33	−0.019 (0.004), 31	−0.032 (0.004), 31
M. margaritifera	−0.087 (0.006), 25	−0.082 (0.008), 19	−0.108 (0.008), 18
Pisidium sp.	0.003 (0.003), 29	0.013 (0.003), 35	0.007 (0.003), 21

Table 3. Parameter estimates for tested models describing the relationship between net calcification rate and pCO_2. Linear models took the form net calcification rate = b0 + b1 ∗ pCO_2. Parabolic models took the form net calcification rate = b0 + b1 ∗ pCO_2 + b2 ∗ $(pCO_2)^2$. Exponential models took the form net calcification rate = b0 + b1 ∗ e^{b2*pCO_2}. Bold rows indicate statistically significant model (α = 0.05) with the lowest AIC value for a given species.

Species	Model	b0	b1	b2	R2	RMSE	F-Statistic	p-Value	AIC
P. leniusculus	**linear**	$\mathbf{-1.79 \times 10^{-3}}$	$\mathbf{-5.42 \times 10^{-6}}$		**0.257**	**0.008**	**10.71 (1, 31)**	**0.0026**	**−217.18**
P. leniusculus	exponential	-1.10×10^{-4}	-4.71×10^{-3}	4.41×10^{-4}	0.260	0.008	5.29 (2, 30)	0.0104	−215.33
P. leniusculus	parabolic	-5.84×10^{-3}	-4.32×10^{-7}	-1.19×10^{-9}	0.260	0.008	5.27 (2, 30)	0.0105	−215.32
C. fluminea	**linear**	$\mathbf{-1.35 \times 10^{-2}}$	$\mathbf{-5.76 \times 10^{-6}}$		**0.052**	**0.024**	**5.13 (1, 93)**	**0.0258**	**−433.29**
C. fluminea	exponential	1.32×10^{-2}	-2.80×10^{-2}	1.49×10^{-4}	0.050	0.024	2.4 (2, 92)	0.0963	−431.04
C. fluminea	parabolic	2.53×10^{-3}	-2.80×10^{-5}	5.64×10^{-9}	0.067	0.024	3.3 (2, 92)	0.0410	−432.78
M. margaritifera	**linear**	$\mathbf{-7.38 \times 10^{-2}}$	$\mathbf{-1.15 \times 10^{-5}}$		**0.099**	**0.032**	**6.6 (1, 60)**	**0.0126**	**−243.26**
M. margaritifera	exponential	-4.67×10^{-2}	-3.66×10^{-2}	8.29×10^{-5}	0.044	0.033	0.52 (2, 59)	0.5973	−229.62
M. margaritifera	parabolic	-7.86×10^{-2}	-4.40×10^{-6}	-1.90×10^{-9}	0.100	0.032	3.27 (2, 59)	0.0447	−241.29
Pisidium sp.	linear	8.06×10^{-3}	-5.23×10^{-8}		0.000	0.017	0.001 (1, 83)	0.9815	−442.70
Pisidium sp.	exponential	-7.37×10^{-3}	1.54×10^{-2}	-3.09×10^{-6}	0.000	0.017	0.0002 (2, 82)	0.9998	−440.70
***Pisidium* sp.**	**parabolic**	$\mathbf{-1.68 \times 10^{-2}}$	$\mathbf{3.37 \times 10^{-5}}$	$\mathbf{-8.68 \times 10^{-9}}$	**0.076**	**0.017**	**3.393 (2, 82)**	**0.0383**	**−447.46**

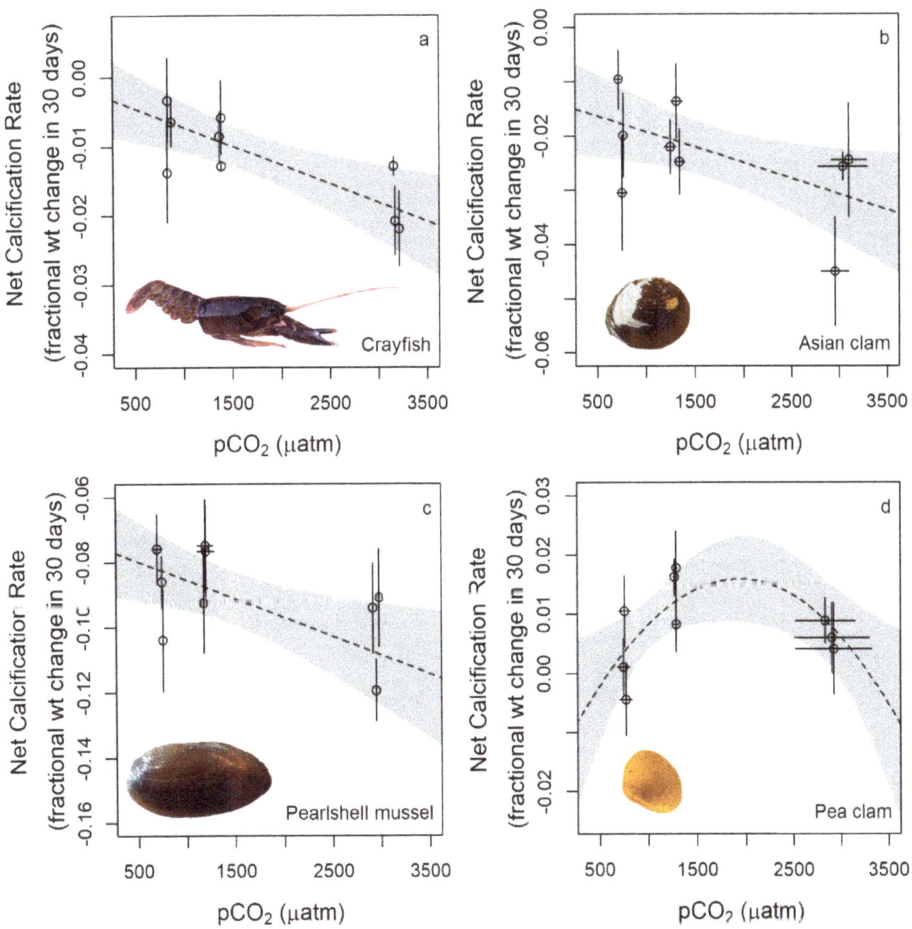

Figure 4. Net calcification rates (expressed as fractional change in buoyant or wet weight normalized

to 30-day growth interval) as a function of water pH for (**a**) the signal crayfish, *Pacifastacus leniusculus* (n = 33, 2–6 individuals per tank); (**b**) the Asian clam, *Corbicula fluminea* (n = 95, 6–15 individuals per tank); (**c**) the Eastern pearlshell mussel, *Margaritifera margaritifera* (n = 62, 2–10 individuals per tank); and (**d**) the pea clam, *Pisidium* sp. (n = 85, 6–15 individuals per tank). All species shown exhibit a linear negative response in net calcification rate to increasing pCO_2, except for the pea clam that exhibits a parabolic response. Data markers represent the average net calcification rates for all individuals in each of the three replicate tanks for each of the three pCO_2 treatments. Vertical bars represent the standard error of the net calcification rates, while the horizontal bars represent the standard deviation of pCO_2 during the incubations. Shaded regions represent the 95% confidence intervals of the best fitting model.

Metabolism and food consumption of the crayfish, *Pacifasticus leniusculus*, exhibited more complex responses to altered pCO_2 treatments. Respiration rates averaged 14.7 μmol g^{-1} h^{-1} (n = 43 crayfish, 13 in low and high pCO_2 treatments, respectively, 17 in the intermediate pCO_2 treatment), but did not vary with pCO_2 (Figure 5a, $F_{1,41} = 0.281$, $p = 0.60$). Feeding rates, however, were impacted by pCO_2 (Figure 5b, ANOVA, $F_{2,15} = 4.47$, $p = 0.030$), such that feeding rates in the highest pCO_2 trials were significantly less than in the intermediate pCO_2 trials (post-hoc Tukey, $p = 0.043$), and with nearly significantly less than in the lowest pCO_2 trials (post-hoc Tukey, $p = 0.060$).

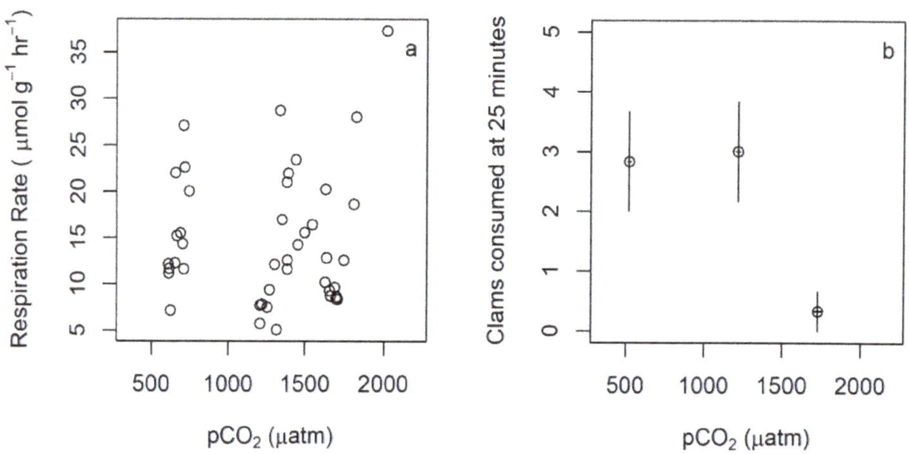

Figure 5. (**a**) Respiration and (**b**) feeding rates of the signal crayfish as a function of pCO_2. High pCO_2 inhibits feedings rates for the crayfish, while respiration rates do not significantly vary across pCO_2 treatments. Vertical bars represent standard error of respiration rates for of feeding rates for six crayfish per treatment; horizontal bars represent standard deviation of pCO_2 for the respective treatments.

4. Discussion

Elevated pCO_2, resulting in decreased pH, caused a reduction in net calcification rates in three species of freshwater calcifiers, the Asian clam *Corbicula fluminea*, the pearlshell mussel *Margaritifera margaritifera*, and the crayfish *Pacifastacus leniusculus*. In the pea clam, *Pisidium* sp., however, net calcification rate was highest under intermediate pCO_2. The linearly negative response of the Asian clam, pearlshell mussel, and crayfish suggests that these species will be negatively impacted by CO_2-induced acidification of freshwater systems over the coming decades. However, optimization of calcification under moderate acidification exhibited by the pea clam suggests that this species may be more resilient to CO_2-induced acidification predicted for the near future, but this resilience may diminish at higher pCO_2 conditions predicted for the next century or two. Although the high inter-specimen variability in growth rates of the pea clam may argue for selection of a simpler model (i.e., linear), the large number of individuals employed in the pea clam experiment (85) supports adoption of the more complex parabolic model.

There are various mechanisms that can lead to the observed functional relationships between net calcification rate and pCO_2. Because we did not manipulate calcium ion concentrations, shell formation here is likely the culmination of two main processes the uptake of inorganic carbon, primarily in the form of bicarbonate ion and, to a lesser extent, aqueous CO_2, and the efflux of protons (H^+) from the site of calcification that effectively converts bicarbonate ion to carbonate ion available for calcification [18,19]. The linear reduction in net calcification rate with decreased pH (increased pCO_2) exhibited by the Asian clam, pearlshell mussel, and crayfish suggests that these species are limited in their ability to control carbonate chemistry at the site of calcification in support of calcification, such as by removing H^+ from their calcifying fluid.

This linear negative calcification response to elevated pCO_2 exhibited by the crayfish is somewhat surprising given that various species of marine decapod crustacea are reported to calcify faster under elevated pCO_2 [5,20], suggesting that freshwater crustacea utilize dissolved inorganic carbon in shell formation differently than marine crustacea and/or that other processes within their physiological repertoire are more sensitive to pH (see discussion below). The parabolic relationship between net calcification rate and pH exhibited by the pea clam suggests that it may be able to utilize the additional dissolved inorganic carbon (DIC) resulting from increased pCO_2 by, for example, converting it to carbonate ions by removing protons from the site of calcification. Under the highest pCO_2 condition, it is possible that the benefits of this process are overwhelmed by other processes that are more deleteriously impacted by the low pH conditions, such as the dissolution of pre-formed, exposed shell in water that is relatively undersaturated with respect to the clam's aragonite [21] shell mineral. This trend may be further reinforced by nonlinearities imposed by scaling differences between surface- vs. volume-related processes. For example, because *Pisidium* sp. broods its larvae [22], calcification occurs throughout the clam's body cavity as its offspring build their shells. This contrasts the calcification processes of *M. margaritifera* and *C. fluminea*, which do not brood their calcifying offspring. Dissolution, however, acts only on shell surfaces exposed to water with a low saturation state, meaning that the brooding larval *Pisidium* sp. are potentially shielded from this process, which may confer some resilience to calcification of the whole pea clam (i.e., both adults and brooded larvae) under intermediate pCO_2 conditions, potentially resulting in their observed parabolic responses to elevated pCO_2.

It is also possible that decreased water pH impacts rates of shell formation indirectly, potentially through physiological processes beyond biocalcification. For the crayfish, *P. leniusculus*, we observed that feeding rates were reduced at the lowest pH despite no pH-dependent variation in respiration rates. We did not test for specific mechanisms, but similar reductions in feeding rates, arising from modified behavior, have been observed for marine decapods exposed to increased pCO_2 [23]. Alternatively, lowered pH has been shown to cause a biomechanical weakening of the exoskeleton in some species of marine crustacea [24], and it is possible that this could impair their ability to handle and consume prey. Regardless of the cause, these reduced feeding rates under CO_2-induced reductions in water pH would yield less food and energy for shell production by the crayfish, potentially contributing to or driving their linear decline in net calcification rate with increasing pCO_2. Although the feeding rates of the other species were not quantified in the present study, obtaining these types of measurements in future experiments should provide valuable insight into the mechanisms responsible for the observed reductions in net calcification rate of freshwater calcifiers under conditions of elevated pCO_2.

Anthropogenic increases in atmospheric pCO_2 and the resulting decrease in freshwater pH has the potential to negatively impact freshwater aquatic ecosystems by altering the net calcification rate and behavior of calcifying invertebrates [1]. Differential sensitivities to CO_2-induced changes in water chemistry can further shape populations, communities, and ecosystems [25]. Given the ecological importance of freshwater calcifiers in lakes, ponds, rivers, and streams [26–28], further research is merited to determine the range of responses that such species exhibit to elevated atmospheric pCO_2, and the mechanisms driving these diverse responses. A better understanding of how freshwater species perform under acidification can help inform strategies for managing freshwater systems amidst the various threats that they face, including species invasion, eutrophication, sedimentation, warming, and CO_2-induced acidification.

5. Conclusions

1. CO_2-induced acidification (doubling, quadrupling of ambient pCO_2 causing pH decline of between 0.23 and 0.57 units) caused negative linear calcification responses in three species of freshwater calcifiers and a more complex parabolic calcification response in one species (pea clam). Increased pCO_2 decreased the feeding rates of the signal crayfish on its natural pea clam prey but had no significant impact on crayfish respiration rate (i.e., constant metabolism), potentially driving this species' negative calcification response to acidification by depriving it of energy required for its molt-mediated calcification.
2. Although increased pCO_2 can impair calcification rates of freshwater organisms, variation in these effects amongst species, combined with impacts on predator–prey dynamics, could yield complex ecological consequences for freshwater systems.
3. Results highlight the importance of further elucidating the understudied effects of CO_2-induced acidification within freshwater systems.

Author Contributions: Conceptualization, A.T.N. and J.R.; data curation, A.T.N. and J.R.; formal analysis, A.T.N. and J.R.; funding acquisition, A.T.N. and J.R.; investigation, A.T.N. and J.R.; methodology, A.T.N. and J.R.; resources, A.T.N. and J.R.; writing—original draft, A.T.N. and J.R.; writing—review and editing, A.T.N. and J.R. All authors have read and agreed to the published version of the manuscript.

Funding: A.T.N. was supported by a Russel J. and Dorothy S. Bilinski Fellowship at Bodega Marine Laboratory. J.R. and this research were supported by MIT Sea Grant award no. NA18OAR4170105, by Northeastern's Interdisciplinary Sabbatical Program, by J.R.'s overhead return fund at Northeastern, and by J.R.'s personal funds. J.R. was also supported by the University of California at Davis Tahoe Environmental Research Center, which provided space, equipment, analytical resources, utilities, and boat/dive time in support of this research.

Institutional Review Board Statement: Not applicable.

Informed Consent Statement: Not applicable.

Data Availability Statement: Data available upon request from the authors.

Acknowledgments: We are grateful to numerous scientists and staff members at the UC Davis Tahoe Environmental Research Center (TERC), including Anne Liston, Brant Allen, Katie Senft, and Brandon Berry, for their assistance in the laboratory and with collections in the field, and to TERC Director Geoffrey Schladow for hosting J.R.'s sabbatical and A.T.N.'s visiting appointment and for generously providing resources in support of this work. We are also grateful to Brian Gaylord for advice and comments on previous versions of the manuscript and to Cary and Cindy Ninokawa for their assistance with crayfish collection and feeding trials.

Conflicts of Interest: The authors are not aware of any conflict of interest. Additionally, the external funders had no role in the design of the study; in the collection, analyses, or interpretation of data; in the writing of the manuscript; or in the decision to publish the results.

References

1. Doney, S.C.; Busch, D.S.; Cooley, S.R.; Kroeker, K.J. The impacts of ocean acidification on marine ecosystems and reliant human communities. *Annu. Rev. Environ. Resour.* **2020**, *45*, 1–30. [CrossRef]
2. Hurd, C.L.; Beardall, J.; Comeau, S.; Cornwall, C.E.; Havenhand, J.N.; Munday, P.L.; Parker, L.M.; Raven, J.A.; McGraw, C.M. Ocean acidification as a multiple driver: How interactions between changing seawater carbonate parameters affect marine life. *Mar. Freshw. Res.* **2020**, *71*, 263–274. [CrossRef]

3. Comeau, S.; Cornwall, C.E.; DeCarlo, T.M.; Krieger, E.; McCulloch, M.T. Similar controls on calcification under ocean acidification across unrelated coral reef taxa. *Glob. Chang. Biol.* **2018**, *24*, 4857–4868. [CrossRef]
4. Gilbert, P.U.P.A.; Bergmann, K.D.; Boekelheide, N.; Tambutté, S.; Mass, T.; Marin, F.; Adkins, J.F.; Erez, J.; Gilbert, B.; Knutson, V.; et al. Biomineralization: Integrating mechanism and evolutionary history. *Sci. Adv.* **2022**, *8*, eabl9653. [CrossRef]
5. Ries, J.B.; Cohen, A.L.; McCorkle, D.C. Marine calcifiers exhibit mixed responses to CO_2-induced ocean acidification. *Geology* **2009**, *37*, 1131–1134. [CrossRef]
6. Gaylord, B.; Barclay, K.M.; Jellison, B.M.; Jurgens, L.J.; Ninokawa, A.T.; Rivest, E.B.; Leighton, L.R. Ocean change within shoreline communities: From biomechanics to behaviour and beyond. *Conserv. Physiol.* **2019**, *7*, 1–14. [CrossRef]
7. George, M.N.; O'Donnell, M.J.; Concodello, M.; Carrington, E. Mussels repair shell damage despite limitations imposed by ocean acidification. *J. Mar. Sci. Eng.* **2022**, *10*, 359. [CrossRef]
8. Waldbusser, G.G.; Gray, M.W.; Hales, B.; Langdon, C.J.; Haley, B.A.; Gimenez, I.; Smith, S.R.; Brunner, E.L.; Hutchinson, G. Slow shell building, a possible trait for resistance to the effects of acute ocean acidification. *Limnol. Oceanogr.* **2016**, *61*, 1969–1983. [CrossRef]
9. Jokiel, P.L. Ocean acidification and control of reef coral calcification by boundary layer limitation of proton flux. *Bull. Mar. Sci.* **2011**, *87*, 639–657. [CrossRef]
10. Hasler, C.T.; Butman, D.; Jeffrey, J.D.; Suski, C.D. Freshwater biota and rising pCO_2? *Ecol. Lett.* **2016**, *19*, 98–108. [CrossRef] [PubMed]
11. Aufdenkampe, A.K.; Mayorga, E.; Raymond, P.A.; Melack, J.M.; Doney, S.C.; Alin, S.R.; Aalto, R.E.; Yoo, K. Riverine coupling of biogeochemical cycles between land, oceans, and atmosphere. *Front. Ecol. Environ.* **2011**, *9*, 53–60. [CrossRef]
12. Hasler, C.T.; Jeffrey, J.D.; Schneider, E.V.C.; Hannan, K.D.; Tix, J.A.; Suski, C.D. Biological consequences of weak acidification caused by elevated carbon dioxide in freshwater ecosystems. *Hydrobiologia* **2018**, *806*, 1–12. [CrossRef]
13. Caldeira, K.; Wickett, M. Ocean model predictions of chemistry changes from carbon dioxide emissions to the atmosphere and ocean. *J. Geophys. Res. C Ocean.* **2005**, *110*, 1–12. [CrossRef]
14. Waters, J.; Millero, F.J.; Woosley, R.J. Corrigendum to "The free proton concentration scale for seawater pH", [MARCHE: 149 (2013) 8-22]. *Mar. Chem.* **2014**, *165*, 66–67. [CrossRef]
15. Nathenson, M. *Chemistry of Lake Tahoe, California-Nevada, and Nearby Springs*; Department of the Interior, US Geological Survey: Reston, VA, USA, 1989. [CrossRef]
16. Jokiel, P. Coral growth: Buoyant weight. In *Coral Reefs: Research Methods*; Stoddart, D.R., Johannes, R.E., Eds.; UNESCO: Paris, France, 1978; pp. 529–541.
17. Shechter, A.; Berman, A.; Singer, A.; Freiman, A.; Grinstein, M.; Erez, J.; Aflalo, E.D.; Sagi, A. Reciprocal changes in calcification of the gastrolith and cuticle during the molt cycle of the red claw crayfish *Cherax quadricarinatus*. *Biol. Bull.* **2008**, *214*, 122–134. [CrossRef] [PubMed]
18. Bach, L.T. Reconsidering the role of carbonate ion concentration in calcification by marine organisms. *Biogeosciences* **2015**, *12*, 4939–4951. [CrossRef]
19. Liu, Y.W.; Sutton, J.N.; Ries, J.B.; Eagle, R.A. Regulation of calcification site pH is a polyphyletic but not always governing response to ocean acidification. *Sci. Adv.* **2020**, *6*, eaax1314. [CrossRef]
20. Kroeker, K.J.; Kordas, R.L.; Crim, R.N.; Singh, G.G. Meta-analysis reveals negative yet variable effects of ocean acidification on marine organisms. *Ecol. Lett.* **2010**, *13*, 1419–1434. [CrossRef]
21. Cummings, K.S.; Graf, D.L. Mollusca: Bivalvia. In *Ecology and Classification of North American Freshwater Invertebrates*; Thorp, J.H., Covich, A.P., Eds.; Elsevier: London, UK, 2010; pp. 309–384. ISBN 978-0-12-374855-3.
22. Guralnick, R. Life-history patterns in the brooding freshwater bivalve Pisidium (*Sphaeriidae*). *J. Molluscan Stud.* **2004**, *70*, 341–351. [CrossRef]
23. Dodd, L.F.; Grabowski, J.H.; Piehler, M.F.; Westfield, I.; Ries, J.B. Ocean acidification impairs crab foraging behavior. *Proc. R. Soc. B Biol. Sci.* **2015**, *282*, 1–9. [CrossRef]
24. Dickinson, G.H.; Bejerano, S.; Salvador, T.; Makdisi, C.; Patel, S.; Long, W.C.; Swiney, K.M.; Foy, R.J.; Steffel, B.V.; Smith, K.E.; et al. Ocean acidification alters properties of the exoskeleton in adult Tanner crabs, *Chionoecetes bairdi*. *J. Exp. Biol.* **2021**, *224*, jeb232819. [CrossRef] [PubMed]
25. Kroeker, K.J.; Micheli, F.; Gambi, M.C. Ocean acidification causes ecosystem shifts via altered competitive interactions. *Nat. Clim. Chang.* **2013**, *3*, 156–159. [CrossRef]
26. Dodds, W.K.; Perkin, J.S.; Gerken, J.E. Human impact on freshwater ecosystem services: A global perspective. *Environ. Sci. Technol.* **2013**, *47*, 9061–9068. [CrossRef] [PubMed]
27. Vaughn, C.C.; Hakenkamp, C.C. The functional role of burrowing bivalves in freshwater ecosystems. *Freshw. Biol.* **2001**, *46*, 1431–1446. [CrossRef]
28. Reynolds, J.; Souty-Grosset, C.; Richardson, A. Ecological roles of crayfish in freshwater and terrestrial habitats. *Freshw. Crayfish* **2013**, *19*, 197–218. [CrossRef]

Journal of
Marine Science and Engineering

MDPI

Article

Physicochemical Control of Caribbean Coral Calcification Linked to Host and Symbiont Responses to Varying pCO_2 and Temperature

Robert A. Eagle [1,2,*], Maxence Guillermic [1,2,3], Illian De Corte [1,2], Blanca Alvarez Caraveo [1], Colleen B. Bove [4,5], Sambuddha Misra [6,7], Louise P. Cameron [8], Karl D. Castillo [4,5] and Justin B. Ries [8]

1 Department of Atmospheric and Oceanic Sciences, Institute of the Environment and Sustainability, Center for Diverse Leadership in Science, University of California, 520 Portola Plaza, Los Angeles, CA 90095, USA
2 Université de Brest Occidentale, Institut Universitaire Européen de la Mer, LGO, Rue Dumont d'Urville, 29280 Plouzané, France
3 Department of Earth, Planetary, and Space Sciences, University of California, Los Angeles, 595 Charles Young Drive E., Los Angeles, CA 90095, USA
4 Environment, Ecology, and Energy Program, University of North Carolina at Chapel Hill, Chapel Hill, NC 27599, USA
5 Department of Earth, Marine, and Environmental Sciences, University of North Carolina at Chapel Hill, Chapel Hill, NC 27599, USA
6 Indian Institute of Science, Centre for Earth Sciences, Malleshwaram, Bengaluru 560012, Karnataka, India
7 The Godwin Laboratory for Palaeoclimate Research, Department of Earth Sciences, University of Cambridge, Cambridge CB2 3EQ, UK
8 Department of Marine and Environmental Sciences, Marine Science Center, Northeastern University, 430 Nahant Rd, Nahant, MA 01908, USA
* Correspondence: robeagle@g.ucla.edu

Citation: Eagle, R.A.; Guillermic, M.; De Corte, I.; Alvarez Caraveo, B.; Bove, C.B.; Misra, S.; Cameron, L.P.; Castillo, K.D.; Ries, J.B. Physicochemical Control of Caribbean Coral Calcification Linked to Host and Symbiont Responses to Varying pCO_2 and Temperature. *J. Mar. Sci. Eng.* **2022**, *10*, 1075. https://doi.org/10.3390/jmse10081075

Academic Editor: Tom Spencer

Received: 13 January 2022
Accepted: 25 July 2022
Published: 5 August 2022

Publisher's Note: MDPI stays neutral with regard to jurisdictional claims in published maps and institutional affiliations.

Abstract: It is thought that the active physiological regulation of the chemistry of a parent fluid is an important process in the biomineralization of scleractinian corals. Biological regulation of calcification fluid pH (pH_{CF}) and other carbonate chemistry parameters ($[CO_3^{2-}]_{CF}$, DIC_{CF}, and Ω_{CF}) may be challenged by CO_2 driven acidification and temperature. Here, we examine the combined influence of changing temperature and CO_2 on calcifying fluid regulation in four common Caribbean coral species—*Porites astreoides*, *Pseudodiploria strigosa*, *Undaria tenuifolia*, and *Siderastrea siderea*. We utilize skeletal boron geochemistry (B/Ca and $\delta^{11}B$) to probe the pH_{CF}, $[CO_3^{2-}]_{CF}$, and DIC_{CF} regulation in these corals, and $\delta^{13}C$ to track changes in the sources of carbon for calcification. Temperature was found to not influence pH_{CF} regulation across all pCO_2 treatments in these corals, in contrast to recent studies on Indo-Pacific pocilloporid corals. We find that $[DIC]_{CF}$ is significantly lower at higher temperatures in all the corals, and that the higher temperature was associated with depletion of host energy reserves, suggesting $[DIC]_{CF}$ reductions may result from reduced input of respired CO_2 to the DIC pool for calcification. In addition, $\delta^{13}C$ data suggest that under high temperature and CO_2 conditions, algal symbiont photosynthesis continues to influence the calcification pool and is associated with low $[DIC]_{CF}$ in *P. strigosa* and *P. astreoides*. In *P. astreoides* this effect is also associated with an increase in chlorophyll a concentration in coral tissues at higher temperatures. These observations collectively support the assertion that physicochemical control over coral calcifying fluid chemistry is coupled to host and symbiont physiological responses to environmental change, and reveals interspecific differences in the extent and nature of this coupling.

Keywords: Caribbean; coral; calcification; pH regulation; boron isotopes; B/Ca; carbon isotopes; photosynthesis; bleaching; symbiont

1. Introduction

Corals are foundational for some of earth's most biologically productive and diverse ecosystems, many of which are threatened by a combination of rising ocean temperatures, disease, ocean acidification (OA), and other pressures. Ocean warming can cause a breakdown in the symbiotic relationship between corals and their algal symbionts leading to coral "bleaching" and eventual mortality [1] and represents a significant threat to corals [2]. In addition, as atmospheric carbon dioxide partial pressure continues to rise, so does uptake of CO_2 into seawater, which reduces seawater pH (pH_{SW}) and calcium carbonate saturation state (Ω), thereby making seawater less chemically favorable for organisms that produce $CaCO_3$ shells and skeletons [3]. However, it is understood that some species of marine calcifying organisms are resilient to external acidification or may even benefit by utilizing the additional dissolved inorganic carbon (DIC) available for shell building, and in some cases photosynthesis [4–7].

Coral calcification responses to ocean acidification are known to be diverse, with observations of negative skeletal growth responses [8–20]. Coupled ocean acidification and temperature stress has revealed a variety of responses due to interactions with temperature stress, including negative synergistic [14,21–24].

Whilst interactive effects of temperature and acidification on corals are known, the underlying physiological causes of those effects are less well understood. In two Indo-Pacific pocilloporid coral species, *Pocillopora damicornis* and *Stylophora pistillata*, we observed that corals cultured at 28 °C exhibit increased net calcification with increasing pCO_2, but a shift to negligible or negative net calcification at 31 °C [7]. At the 28 °C treatment, CO_2 fertilization of symbiont photosynthesis appeared to be occurring, which supplied the energy resources for corals to increase calcification [7,25]. At 31 °C, loss of symbiont density was observed from coral color analyses [7,25]. Using two independent approaches, pH microelectrodes and skeletal boron geochemistry, it was found that temperature compromised the corals' ability to maintain biological regulation of its internal parent fluids for calcification (calcifying fluids; CF), with pH_{CF} and $[CO_3^{2-}]_{CF}$ reduced at 31 °C and more sensitive to changes in the carbonate chemistry of external seawater [7]. Similar observations using pH sensitive dyes [26] and geochemical proxies [27–29] have been reported in field-collected corals with different thermal exposure histories. As it is thought that biological upregulation of the pH_{CF}, $[CO_3^{2-}]_{CF}$, $[DIC]_{CF}$, and the saturation state of aragonite $(\Omega_{Ar})_{CF}$ is important in allowing corals to calcify [30–35], loss of this control on thermal stress may explain how temperature and pCO_2-induced ocean acidification could interact to negatively impact coral calcification [7]. Both laboratory experiments and observations on field collected corals have their limitations. In culture experiments, the question is how well are natural physiological responses to stressors simulated by the experiment and, in field collected corals, it can be difficult to constrain and disentangle complex and covarying relationships between the seawater carbonate system and their geochemical proxies [36]. Whilst coral calcifying chemistry regulation in response to acidification has been widely investigated in Indo-Pacific species, it has not to our knowledge been examined in depth in culture experiments on the Caribbean species studied here, with the exception of a study by [37] on *S. siderea* using a spatially resolved laser ablation approach.

Here, we further explore the independent and combined effect of temperature and pCO_2 on coral and algal symbiont physiology in a study on four species of common Caribbean scleractinian corals *Porites astreoides*, *Pseudodiploria strigosa*, *Undaria tenuifolia*, and *Siderastrea siderea* (Figure 1). These corals were recovered from the Belize Mesoamerican Barrier Reef System, cultured at 28 °C and 31 °C, and at ~280 to 3300 µatm pCO_2, and characterized for their structural and physiological responses to these conditions [38,39]. The coral species exhibited interspecific variations in host and symbiont responses to changing environmental conditions in the experiment, which allows us to investigate the link between host and symbiont physiology and the control of calcification and calcification fluid chemistry [38,39]. It was found that all four coral species exhibited nonlinear decreases in net calcification rate with increasing pCO_2 [38]. The 31 °C treatment reduced calcification rate in only one species, *P. strigosa*, although *U. tenuifolia* exhibited too low survival rates at 31 °C to be assessed [38]. *Siderastrea siderea* was the most resilient coral maintaining positive rates of net calcification under all conditions, even when seawater was undersaturated with respect to aragonite [38]. *Siderastrea siderea* was found to maintain relatively unaltered symbiosis under experimental conditions, and relatively constant host energy reserves, consistent with its resilient calcification response [39]. *Pseudodiploria strigosa* did not maintain indicators of coral symbiont physiology under warming, such as chlorophyll content and symbiont cell density, and was the most bleached coral in the experiment [39]. Conversely, *P. astreoides* exhibited improved symbiont physiological indicators and chlorophyll content at 31 °C, which were reduced under the acidification conditions [39], although the 31 °C condition was characterized by lower survival rates [38]. *Pseudodiploria strigosa* and *S. siderea* showed stronger correlations between symbiont density and host energy reserves, consistent with a tighter coupling between these physiological indicators than were found in *P. astreoides* [39]. It was concluded in previous work that this may result from *P. astreoides* symbionts being more efficient in delivering autotrophically derived carbon to the host, whereas *P. strigosa* and *S. siderea* require relatively greater concentrations of their symbionts to support the host [39]. Therefore, this experiment represents an opportunity to expand on our previous work [7] to explore the underlying mechanisms behind the interaction of temperature and pCO_2 on corals and, in particular, how symbiont and host physiological responses could influence corals' ability to exert active biological control on physicochemical parameters, such as pH_{CF}, $[DIC]_{CF}$, and $[CO_3^{2-}]_{CF}$. We applied a combined approach of boron geochemistry to constrain pH_{CF}, $[DIC]_{CF}$, and $[CO_3^{2-}]_{CF}$ and $\delta^{13}C$ to trace changes in the source of carbon for calcification linked to coral and symbiont physiological responses to environmental changes.

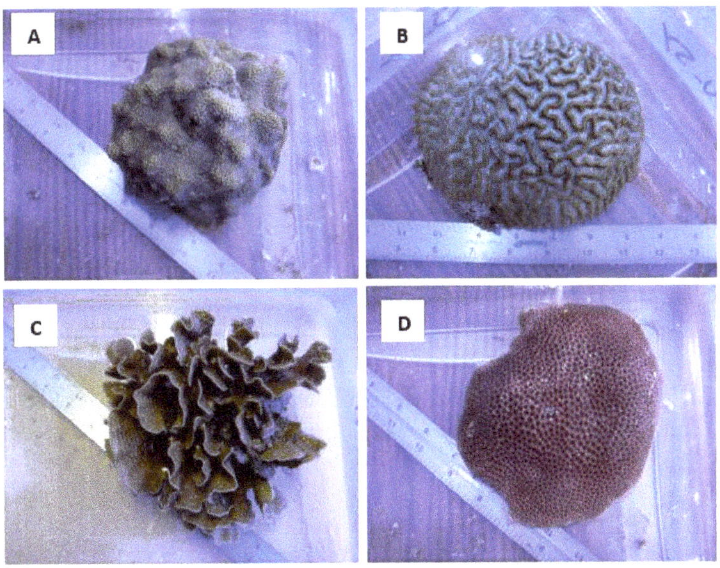

Figure 1. Sample photographs of the coral species cultured in the experiment of Bove et al. (2019) [38] and investigated here: (**A**) *Porites astreoides*, (**B**) *Pseudodiploria strigosa*, (**C**) *Undaria tenuifolia*, and (**D**) *Siderastrea siderea*.

2. Materials and Methods

2.1. Sample Collection

Samples from the tropical Caribbean Scleractinian coral species *Siderastrea siderea*, *Pseudodiploria strigosa*, *Porites astreoides*, and *Undaria tenuifolia* (Figure 1) were collected from inshore and offshore reef environments along the southern portion of the Mesoamerican

Barrier Reef System (MBRS) off the coast of Belize in June 2015, at depths of 3–5 m [38]. For each species, six colonies were collected from both an inshore and offshore reef environment, totaling 48 colonies (4 species × 6 colonies × 2 reef environments), as described in Supplementary Table S3. Samples were then transported to a natural seawater flow-through aquarium system at Northeastern University's Marine Science Center in Nahant, MA, USA, where each colony was sectioned into eight fragments to be acclimated and subsequently undergo the 93-day experiment [38]. New data from a second experiment on Indo-Pacific corals are also presented here, in this experiment specimens of *Stylophora pistillata*, were obtained from DeJong MarineLife (Lingewaal, The Netherlands) and cultured as described previously [7,25].

2.2. Experimental Culturing

Following a ~70 day conditioning period, fragmented coral colonies from *S. siderea*, *P. strigosa*, *P. astreoides*, and *U. tenuifolia* were reared for 93 days under four pCO_2 conditions corresponding to preindustrial (311 μatm), present-day control (405 μatm), end-of-century (701 μatm), and extreme (3309 μatm) carbon dioxide partial pressures, and under two temperature conditions corresponding to control (28 °C, as determined by in situ reef temperature records near the collection site) and elevated temperature treatments (31 °C), as predicted for end-of-century mean sea-surface temperature near the collection site [40]. The extreme pCO_2 condition allows for an exploration of the physiological limits of the species, as well as simulates conditions in past greenhouse periods in earth history which are relevant to understand coral evolution. Coral fragments within each tank were fed every other day with a mixture of ca. 6 g frozen adult *Artemia* sp. and 250 mL concentrated newly hatched live *Artemia* sp. (500 mL^{-1}) to satisfy any heterotrophic feeding by each species. Supplementary Tables S1 and S2 reproduce the measured and calculated experimental parameters from the original study [38].

2.3. Net Calcification Rate, Coral Symbiont and Host Physiological Measurements

Net coral calcification rates from this experiment were estimated from surviving coral fragments using the buoyant weight method, as described in [38], and reported in units of mg cm^{-2} d^{-1}. Algal symbiont cell density was assessed using Lugol's iodine staining, with cell counts standardized to total coral tissue volume to yield values in cells per cm^2 as described in [39]. Chlorophyll a (Chl a) was measured on a Turner Design 10-AU fluorometer and expressed in units of g of pigment per cm^2 of coral tissue surface area [39]. Coral host protein, host lipids, and host carbohydrate were determined by spectrophotometry and recorded in units of mg per cm^2 of coral tissue surface area. Whilst these data have been reported in previous publications [38,39], they are reproduced here in Table S3 to provide easy reference for interpreting the geochemical data.

2.4. Coral Sample Preparation for Geochemical Analyses

New skeletal growth was identified relative to a calcein marker emplaced in the coral skeleton at the start of the experiment, which was visualized using fluorescent light microscopy [38]. Powdered and homogenized samples underwent clay-removal and oxidative cleaning following the method from Barker et al. [41]. Sample material was then dissolved in 1 M HCl (80–100 μL).

2.5. Boron Isotopic Analysis

Boron was purified via microdistillation [42,43] following similar methods described in our previous work [7,44]. Measurements were carried out on a Thermo Scientific ®Neptune Plus MC-ICP-MS at the University of Cambridge equipped with 10^{13} Ω resistors [45]. Boron isotopic composition is reported in standard $\delta^{11}B$ per mil (‰) notation with respect to the NIST SRM 951a boric acid standard (Catanzaro et al., 1970). A laboratory coral standard (NEP; *Porites* sp.) from the University of Western Australia and the Australian National

University was analyzed for assessing internal reproducibility between analytical sessions as well as external reproducibility with other labs.

Sample blanks typically contained less than 0.6 ng-B (<5 ppb B). The δ^{11}B composition of the NEP ($\delta^{11}B_{NEP}$) was measured at 25.6 ± 1.0‰ (2 SD, n = 20) across 13 analytical sessions (Table S4), with each number representing an ab initio processed sample from the present study, which are within analytical uncertainty of published values for the same in-house standard of 25.96 ± 0.30‰, 26.2 ± 0.88‰, n = 27, (25.8 ± 0.89‰, n = 6, 25.71 ± 0.79‰, n = 27, from [46–48]. The proposed international standard JCP−1 [49] was not analyzed across our analytical sessions due to lack of availability, nevertheless, previous work [7,44] was performed over the same period and report values within analytical uncertainty of published data, thereby validating the method utilized in the present study [7,49].

2.6. B/Ca Analysis

Elemental ratios were analyzed on a Thermo Scientific® Element HR ICP-MS at Cambridge University, UK and at European Institute for Marine Studies (IUEM), France, after calcium concentration checks on the Agilent® ICP-OES at Cambridge University, UK and at IUEM, France. Isotope concentrations of the acid blanks relative to a typical Ca concentration session (10 ppm) were: Li^7 < 7%, B^{11} < 0.75%, Mg^{25} < 0.5%, Sr^{87} < 0.01%, Ca^{43} < 0.1%. External reproducibility is reported relative to the consistency standard CamWuellestorfi, as published by [43] and measurements of the NEP coral. Our data were within one standard error of published values (Table S5). The analytical uncertainties (2sd) on the B/Ca ratios were 15 µmol/mol.

2.7. Modelled pH_{CF}

The pH of the calcifying fluid (pH_{CF}) was calculated from the boron isotopic composition of the coral skeleton following rationale from Hemming and Hanson [50] and using the MATLAB code provided in De Carlo et al. [51], where $\delta^{11}B_{SW}$ represents the boron isotopic composition of seawater ($\delta^{11}B_{SW}$ = 39.61‰) [52] and $\delta^{11}B_{CARB}$ represents the boron isotopic composition of the coral skeleton. The isotopic fractionation factor between the boron forms from Klochko et al. (2006) was used, $\alpha_{(B3\text{-}B4)} = 1.0272$. The dissociation constant of boron (K_B) was calculated from salinity, temperature and pressure [53], and the boron concentration was calculated from salinity after Lee et al., 2010 [54].

$$pH_{CF} = pK_B - \log\{-[\delta^{11}B_{SW} - \delta^{11}B_{CARB}]/[\delta^{11}B_{SW} - \alpha_{(B3\text{-}B4)}{*}\delta^{11}B_{CARB} - \varepsilon_{(B3\text{-}B4)})]\} \quad (1)$$

$[H^+]_{SW}$ and $[H^+]_{CF}$ were calculated based on pH_{SW} and pH_{CF} with $[H^+] = 10^{-pH}$. The proton gradient $[H^+]_{SW} - [H^+]_{CF}$ was evaluated to better visualize the scale of pH regulation in comparison to a logarithmic scale.

2.8. Modelled $[CO_3^{2-}]_{CF}$, DIC_{CF}, and Ω_{CF}

DIC_{CF} and $[CO_3^{2-}]_{CF}$ were calculated using the method and Matlab© code from DeCarlo et al. [51]. This study uses combined B/Ca and δ^{11}B to calculate $[CO_3^{2-}]_{CF}$, where:

$$\left[CO_3^{2-}]_{CF}\right] = K_D * \frac{[B(OH)_4^-)]}{B/Ca} \quad (2)$$

We use the KD defined in McCulloch et al. [33], KD = 0.00297*exp(−0.0202*[H$^+$]), where [H$^+$] is determined based on pH_{CF} as calculated from δ^{11}B following Equation (1). Given pH_{CF} and $[CO_3^{2-}]_{CF}$, DIC_{CF} can be calculated using the known dissociation constants for carbonate speciation at known temperatures and pressure. Ω_{CF} can also be calculated but requires and additional assumption on the [Ca^{2+}] of the calcifying fluid. The saturation state of aragonite in the calcifying medium (Ω_{CF}) defined following Equation (3) is dependent on the accuracy of the $[CO_3{}^{2-}]_{CF}$ and is limited to the additional assumption on the $[Ca^{2+}]$ of the calcifying fluid.

$$[\Omega]_{CF} = \frac{\left[CO_3^{2-}]_{CF}\right] \times \left[Ca^{2+}]_{CF}\right]}{Ksp} \tag{3}$$

We calculated the saturation state of aragonite of the calcifying fluid using the $[Ca^{2+}]_{sw}$ determined using the salinity of the culture condition ($[Ca^{2+}]_{sw}$ (mol/kg sw) = 0.02128/40.1* ((Salinity − 0.03)/1.805)).

2.9. Carbon Stable Isotope Analyses

Analyses on Caribbean corals were conducted on two Nu Perspective isotope ratio mass spectrometers (IRMS) at UCLA [55]. Most samples were analyzed using a Nu Carb specific sample preparation system. This system reacts 0.48 mg (±0.03 mg) of pure calcium carbonate material, for 20 min at 70 °C in individual reaction vials, thus eliminating any potential memory effects that are associated with analyses using a common acid bath system. Gases released by acid digestion of $CaCO_3$ are purified in a series of liquid N2-cooled, temperature controlled cold-fingers, an Adsorption Trap (AdTrap), an in-line, short Gas Chromatograph (GC) column packed with Porapak Type-QTM50/80 and silver wool, before introduction to the mass-spectrometer dual inlet for isotope analysis. The second digestion system is the common acid bath (CAB) system [55]. Analyses of *Stylophera pistillata* skeletal $\delta^{13}C$ and $\delta^{18}O$ was carried out on a GasBench II coupled to a Delta V mass spectrometer at the stable isotope facility of Pôle spectrometrie Océan (PSO), Plouzané, France. Results were calibrated to the Vienna Pee Dee Belemnite (V-PDB) scale and referenced to the international standard NBS19.

2.10. Statistical Methods

For elemental data, the most restrictive false discovery rate (Q = 0.1%) was used to thin data using the ROUT method set, meaning that <0.1% of identified outliers would have been falsely identified as such. To test for statistically significant changes in net calcification, we performed a one-way ANOVA followed by Dunnett's multiple comparisons tests on skeletal B/Ca composition, $\delta^{11}B$, $\delta^{13}C$, pH_{CF}, and Ω_{CF} across experimental treatment conditions relative to the control condition (28 °C, 400 μatm pCO_2). To determine the best fit by linear or second-order polynomial (quadratic) regressions, we used a combined linear and quadratic fit test, using the Akaike's Information Criteria (AIC). We compared the best-fit regressions across temperature conditions by first determining if the slopes of the two regressions are statistically different, and if the slopes are not significantly different (p-value > 0.05), we tested to see if the intercepts were significantly different (p-value < 0.05). This was done using an Analysis of Covariance (ANCOVA) method. The statistical analyses above were carried out GraphPad Prism version 9.0.0 for macOS, GraphPad Software, San Diego, CA, USA.

$\delta^{11}B$, B/Ca, and $\delta^{13}C$ data were tested to see if it met assumptions of normality using the Shapiro–Wilks test ($p > 0.05$). At the level $p < 0.05$ we found that most of the data were not normally distributed. In order to perform a two-way AVOVA test, we transformed the data using the Tukey ladder of powers method to reduce negative and positive skew of non-normal data methodologically [56]. Method until it met the test of normality. Table S6 gives information about whether data were transformed and how they were transformed. Following confirmation that the data were normal, we analyzed isotope and elemental data using a two-way ANOVA to test the influence of the individual and combined effects of pCO_2 and temperature. The same approach was taken to analyze the calculated parameters pH_{CF}, $[CO_3{}^{2-}]_{CF}$, and DIC_{CF} and the coral and symbiont physiological data; calcification rate, symbiont cell density, chlorophyll a, and total host energy data from [38,39]. All statistical analysis was conducted using the 'MASS' package in R version 4.0.2 (R Core Development Team 2016, Vienna, Austria).

As coral species were collected from inshore and offshore reef environments, so we also tested whether there were significant effects of temperature, pCO_2, and reef environment on geochemical and physiological parameters in a three-way ANOVA. As this analysis did not resolve a significant effect of reef environments, it justified the pooling of data across reef environments to conduct the two-way ANOVA tests described above.

2.11. Data Compilation

Data from previous publications was compiled to construct synthesis figures on coral calcifying fluid chemistry derived from geochemistry and other approaches such as electrodes and pH sensitive dyes. Literature data used for this metanalysis is summarized in Table S12 and was derived from the following publications [7,17,32,34,35,47,57–70]. In order to include the data from Allison et al., 2014 [32] on the carbonate chemistry of the calcification fluid from *P. damicornis* cultured under different concentrations of the ruthenium red (RR, 3.7 and 5.3 μM) inhibitor of Ca-ATPase activity, new calculations of the original data were performed using the framework described in this paper. Carbonate chemistry of the cultured experiments were calculated from alkalinity and pH provided in a previous publication [71] and using the CO_2sys program [72].

3. Results

3.1. Analyses of ($\delta^{11}B$), B/Ca, and $\delta^{13}C$

Mean boron geochemistry ($\delta^{11}B$ and B/Ca) and $\delta^{13}C$ for each experimental condition of the Caribbean coral experiment are reported in Table 1, individual coral specimen data are reported in Table S7. B/Ca (±SD) measurements for the control culturing conditions were averaged for each species under the experimental control conditions of 28 °C and 405 μatm pCO_2: *P. astreoides* (437 ± 31 μmol/mol), *P. strigosa* (495 ± 70 μmol/mol), *S. siderea* (436 ± 18 μmol/mol), *U. tenuifolia* (518 ± 32 μmol/mol). Under the experimental control culturing conditions, coral skeletal aragonite $\delta^{11}B$ compositions (±SD) were averaged for each species: *P. astreoides* (24.54 ± 0.74‰), *P. strigosa* (24.40 ± 1.40‰), *U. tenuifolia* (24.63 ± 0.68‰), and *S. siderea* (23.87 ± 0.47‰). Mean $\delta^{13}C$ (±SD) for each species under the control temperature and pCO_2 experimental condition (28 °C, 405 μatm pCO_2) was: *P. astreoides* (−2.58 ± 0.92‰), *P. strigosa* (−1.11 ± 1.51‰), *S. siderea* (−3.94 ± 0.61‰), and *U. tenuifolia* (−2.65 ± 0.34‰) (Table 1).

Boron geochemical data for *S. pistillata* have previously been reported in Guillermic et al. Here, we report $\delta^{13}C$ and $\delta^{18}O$ data from *S. pistillata* skeletal aragonite from this additional experiment on Indo-Pacific corals, given in Table S8. The principal reason for presenting these data in this study is that they provide context for the interpretation of data from the Caribbean corals, as described in more detail below.

Trends in measured geochemical parameters as a function of culture seawater pH (pH_{SW}) were explored using Akaike's Information Criteria (AIC) to determine whether a linear or second order polynomial (quadratic) regression best fit the data (Figure 2; Table S9). In all cases, a linear regression was found to fit the data better, with the exception of $\delta^{11}B$ and B/Ca data for *S. siderea*, where a nonlinear fit was preferred (Figure 2, Table S9). In terms of geochemical proxy development for historical seawater pH reconstruction, the three species that exhibited relationships between $\delta^{11}B$ and pH_{SW} that were best described by linear regressions seem best suited for this endeavor. *P. strigosa* showing the greatest sensitivity of $\delta^{11}B$ to pH_{SW}, with a regression slope of 2.15 compared to 1.46 in *P. astreoides* and *U. tenuifolia* at the 28 °C temperature condition. However, *S. siderea* showed the greatest sensitivity with a slope of the linear regression between $\delta^{11}B$ and pH_{SW} of 2.52 but was best fit by a nonlinear regression. At 31 °C, it is notable that the relationship between $\delta^{11}B$ and pH_{SW} within *P. strigosa* was no longer statistically significant, indicating a potential influence of temperature on $\delta^{11}B$ in this species, although this was not significant in the ANOVA tests described below (Figure 2, Tables S9 and S11).

Table 1. Average Geochemical Data per Experimental Condition.

Species	T (°C)	CO_2 (µatm)	$\delta^{11}B$ (‰) [1]	Error (SD)	N *	B/Ca (µmol/mol)	Error (SD)	N	$\delta^{13}C$ (‰) [2]	Error (SD)	N
P. astreoides											
	28	311	23.2	0.9	6	396	54	6	−2.3	0.5	4
	28	405	23.6	0.9	6	437	31	6	−2.6	0.9	4
	28	702	22.7	1.4	5	432	53	4	−5.2	1.5	4
	28	3309	22.1	0.8	6	410	46	6	12.5	3.1	3
	31	288	24.1	0.1	1	389	17	1	−1.4	0.0	1
	31	442	23.4	1.3	5	424	32	5	−2.6	0.7	4
	31	674	23.1	0.4	5	443	82	5	−3.5	1.3	4
	31	3285	22.6	0.5	2	568	51	2	−5.0	1.2	2
P. strigosa											
	28	312	24.2	0.6	6	496	43	6	−2.2	0.6	5
	28	406	24.5	1.5	5	495	70	4	−1.1	1.5	3
	28	702	23.3	0.7	6	481	35	6	−4.0	0.6	4
	28	3320	22.2	1.3	6	455	41	6	10.9	1.6	4
	31	288	23.9	0.9	4	527	44	4	−1.8	0.6	4
	31	443	23.9	0.8	5	495	50	5	−1.3	0.7	4
	31	674	23.7	0.5	4	528	33	4	−3.4	0.4	4
	31	3284	23.0	1.5	4	530	54	2	−6.4	2.9	4
S. siderea											
	28	312	24.5	0.7	6	450	26	6	−3.3	0.2	4
	28	405	23.9	0.5	6	436	18	6	−3.9	0.6	4
	28	703	22.7	0.9	6	424	21	6	−6.2	1.1	4
	28	3317	21.8	1.2	6	415	27	6	12.5	1.7	3
	31	288	24.2	0.4	5	467	11	5	−2.6	0.2	3
	31	449	23.4	0.7	6	425	22	6	−2.6	1.4	4
	31	673	22.1	0.7	5	432	10	4	−5.5	1.4	3
	31	3285	21.3	1.4	6	428	21	6	14.3	2.0	3
U. tenuifolia											
	28	312	24.59	0.68	4	490	51	4	−2.2	0.4	4
	28	404	24.25	0.56	4	518	32	4	−2.6	0.3	4
	28	698	23.54	1.25	4	464	76	4	−3.8	1.0	4
	28	3303	22.53	1.94	4	461	62	4	−7.2	2.8	4

* number of specimens analyzed; [1] NIST; [2] V-PDB.

To further constrain trends in the data, two-way ANOVA tests were used to explore the individual and combined effects of pCO_2 and temperature on geochemical parameters. For *P. astreoides*, pCO_2 alone had a significant individual effect on $\delta^{11}B$ and $\delta^{13}C$, while temperature had no effect on geochemical parameters. However, the combined effect of temperature and pCO_2 was found to significantly impact B/Ca. For *P. strigosa*, pCO_2 had a significant effect on $\delta^{11}B$ and $\delta^{13}C$, and temperature only influenced B/Ca. In *S. siderea*, pCO_2 had a significant effect on $\delta^{11}B$, $\delta^{13}C$, and B/Ca, whereas temperature had an effect on $\delta^{13}C$ alone. Lastly, in *U. tenuifolia*, only pCO_2 had a significant effect on $\delta^{13}C$, and no temperature effects were observed. Results of these analyses are presented in Tables 2 and S10.

Coral specimens were collected from both the inshore and offshore reef environments (Supplementary Table S3). Although previous work showed that reef environment had an effect on the physiological response of the corals [39], we found no significant effect of reef environment on the geochemical data or on the carbonate system parameters calculated for the coral calcifying fluid (Table S10). A lack of effect of local reef environment in skeleton boron and carbon geochemistry is a positive feature in terms of using the species as palaeoceanographic archives.

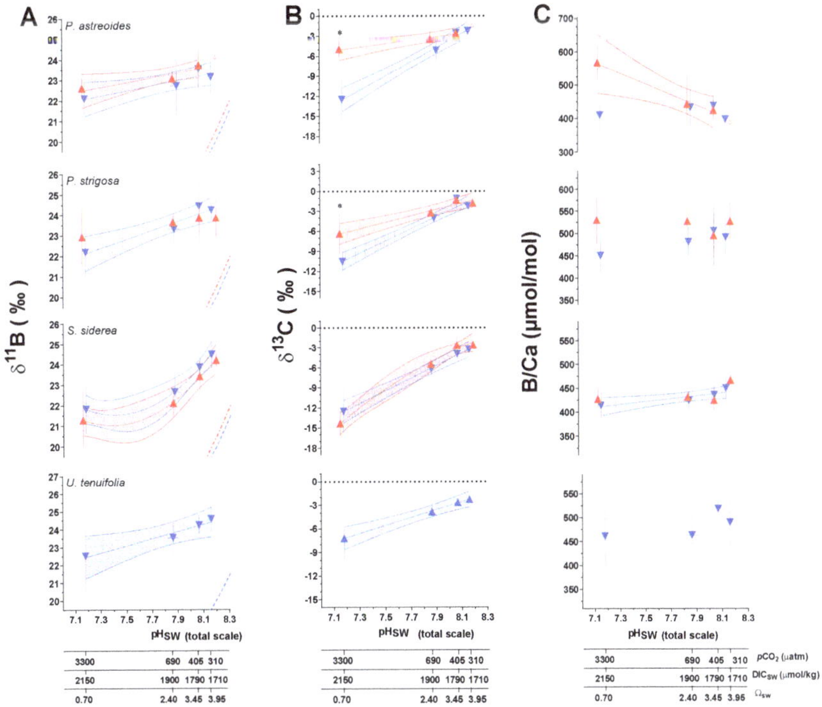

Figure 2. Measured $\delta^{11}B$ (**A**), $\delta^{13}C$ (**B**), and B/Ca (**C**) composition of coral aragonite following the 93-day culturing experiment. Large triangular symbols represent the mean value (±1SD) for each treatment condition. Blue symbols represent the control temperature (28 °C) treatment condition and red represents the high temperature (31 °C) treatment condition. Linear versus centered quadratic fit was determined using the Akaike's Information Criteria test (Table S8), with shaded areas representing the 95% confidence interval. Dashed blue and red curves represent the expected $\delta^{11}B$ composition of borate ion ($B(OH)^{4-}$) in solution at 28 °C and 31 °C, respectively. * indicates a significant difference between 28 °C and 31 °C data for a given pCO_2 treatment as demonstrated by a p-value < 0.05 in a Welch's T test.

It is important to note that the principal driver of trends in $\delta^{13}C$ data across variable pCO_2 and pH_{SW} treatments is the isotopic composition of the CO_2 gas used to manipulate the pCO_2 of the treatment seawater. This gas is combustion sourced and therefore has a relatively negative $\delta^{13}C$ composition that influences the $\delta^{13}C$ of DIC more in cultures with higher pCO_2. Despite these trends being influenced by the $\delta^{13}C$ of the source gas, there is potential to meaningfully interpret the difference between the $\delta^{13}C$ values between the two temperature treatments for a given pCO_2. For example, Welch's T-Tests indicate a statistically significant difference in $\delta^{13}C$ data between the 28 °C and 31 °C treatments at the highest pCO_2 treatment for *P. astreoides* and *P. strigosa*, but not at the lowest pCO_2 condition (Figure 3). Additionally, ANCOVA tests reveal the linear regressions through $\delta^{13}C$ data as a function of pCO_2 for both *P. astreoides* and *P. strigosa* have significantly different slopes at 28 °C versus 31 °C ($p < 0.0001$ and $p = 0.0021$, respectively), while *S. siderea* did not exhibit a significant change in slope across temperature conditions ($p = 0.102$). These trends in $\delta^{13}C$ are likely to reflect changes in the sources of DIC for calcification linked to the coral host and symbiont physiological responses to temperature and pCO_2, as further explored in the following sections.

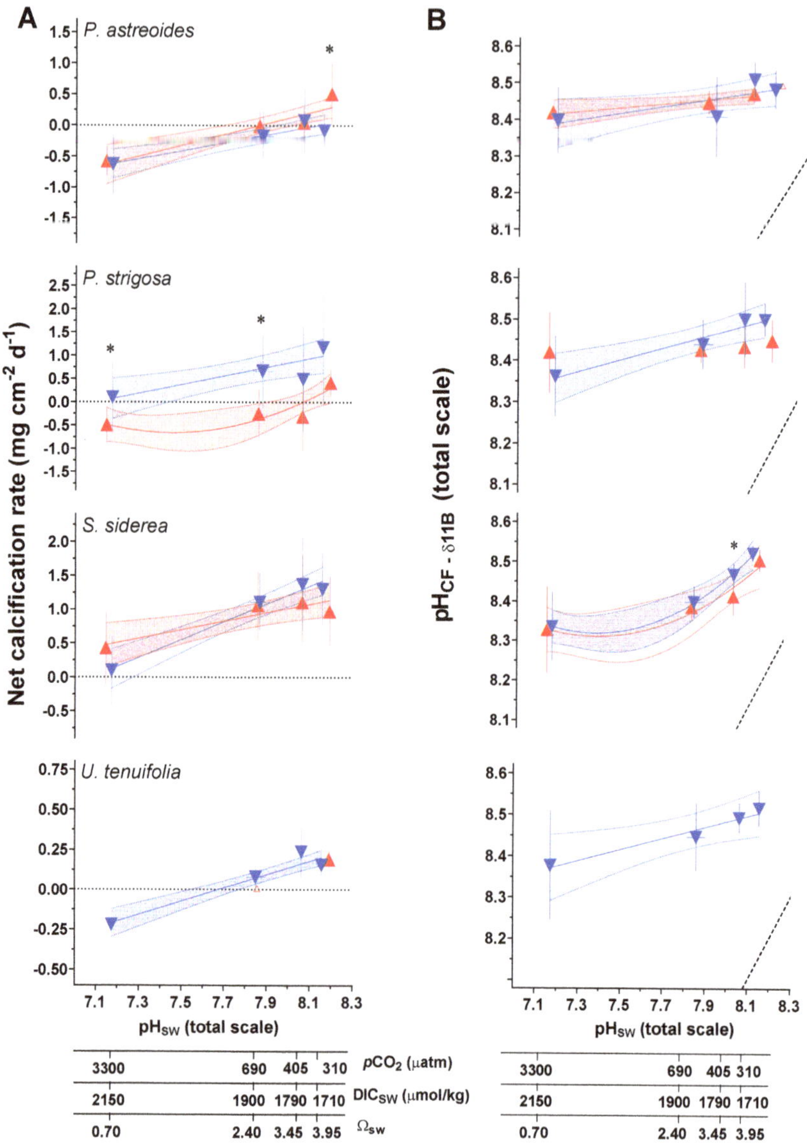

Figure 3. (**A**) Measured net calcification rate (mg cm^{-2} d^{-1}), replotted from Bove et al. [38], (**B**) $\delta^{11}B$-derived pH_{CF}. Triangle symbols represent the mean value (±1 SD) for each treatment condition, blue and red symbols represent the 28 °C and 31 °C treatments, respectively. Linear versus centered quadratic fit was determined using the Akaike's Information Criteria test (Table S8), with shading representing the 95% confidence interval. * indicates a significant difference between 28 °C and 31 °C data for a given pCO_2 treatment, as demonstrated by a p-value < 0.05 in a Welch's T-test.

Table 2. 2-way ANOVA output testing the influence of temperature and pCO_2 on measured geochemical parameters.

Geochemical Parameter	Species	Experimental Variable	ANOVA p-Value
$\delta^{11}B$	*P. astreoides*	Temperature	0.1171
		CO_2	**0.0152**
		Temperature + CO_2	0.6094
	P. strigosa	Temperature	0.5521
		CO_2	**0.00532**
		Temperature + CO_2	0.59757
	S. siderea	Temperature	0.36
		CO_2	**7.58×10^{-10}**
		Temperature + CO_2	0.441
	U. tenuifolia	CO_2	0.322
$\delta^{13}C$	*P. astreoides*	Temperature	0.152
		CO_2	**4.12×10^{-5}**
		Temperature + CO_2	0.193
	P. strigosa	Temperature	0.5521
		CO_2	**0.00532**
		Temperature + CO_2	0.59757
	S. siderea	Temperature	0.36
		CO_2	**7.58×10^{-10}**
		Temperature + CO_2	0.441
	U. tenuifolia	CO_2	0.322
B/Ca	*P. astreoides*	Temperature	0.22201
		CO_2	0.78951
		Temperature + CO_2	**0.00939**
	P. strigosa	Temperature	**0.00911**
		CO_2	0.74462
		Temperature + CO_2	0.16759
	S. siderea	Temperature	0.17474
		CO_2	**0.00103**
		Temperature + CO_2	0.84438
	U. tenuifolia	CO_2	0.484

3.2. Net Calcification, pH_{CF}, $[CO_3^{2-}]_{CF}$, and DIC_{CF}

Net calcification rates for the Caribbean corals were reported in Bove et al. [38], which found that all four coral species showed non-linear reductions in net calcification with increasing pCO_2. Corals showed interspecific differences, with *S. siderea* maintaining positive net calcification under all experimental conditions, *P. strigosa* exhibited a shift to negative net calcification at 31 °C only, and *P. astreoides* exhibited an increase in calcification with elevated temperature at the lower pCO_2 condition, but a shift toward negative calcification (i.e., net dissolution) under the higher pCO_2 treatments (Figure 3).

The pH of the coral calcifying fluid (pH_{CF}) was calculated from $\delta^{11}B$ data using canonical methods from the literature, as described in the methods section. Although the $\delta^{11}B$ approach to calculating pH_{CF} in corals has been independently verified based on comparison with pH sensitive dye and pH microelectrode approaches [7,30,47,61,73], significant differences in sensitivities in those approaches appear to exist, and different hypotheses have been advanced to explain these differences [7,47]. At the control temperature and pCO_2 experimental condition (28 °C, 405 µatm pCO_2), mean $\delta^{11}B$-derived pH_{CF} (total scale, ±1 SD) for each species was as follows: *P. astreoides* (8.45 ± 0.06), *P. strigosa* (8.50 ± 0.11), *S. siderea* (8.47 ± 0.03), *U. tenuifolia* (8.49 ± 0.04) (Table 3).

The $[CO_3^{2-}]_{CF}$ and DIC_{CF} were calculated from $\delta^{11}B$ and B/Ca data using the K_D of McCulloch et al. [74] and the approach of DeCarlo et al. [51] (Table 3), as described in in the methods section. The B/Ca proxy is still prone to uncertainties [70,75] and would benefit from additional validation. One critique is that the proxy is based on limited inorganic precipitation datasets [76,77] which have been difficult to reconcile due to different experimental conditions and from varying $[Ca^{2+}]$ in the experiment leading in turn to varying saturation states of aragonite (Ω) within the precipitation experiments [70]. This leads to potential difficulties when translating to highly biologically controlled fluids (such as the calcifying fluid in corals). Additionally, it is possible that coral biomineralization involves transformation of amorphous precursor phases, as evidenced by the detection of amorphous calcium carbonate (ACC) in corals [78], although it is unclear whether this is the universal pathway by which corals calcify. Amorphous or metastable phases of calcium carbonate may have different K_D, which could contribute to differences in biomineral chemistry [79,80], although given that ACC transformation seems to occur via dissolution and reprecipitation into crystalline phases [81], precursor phase K_D may not be relevant for the final mineral. Despite these uncertainties, results of Guillermic et al. [7] showed that the K_D formulations used by DeCarlo et al. [51] (Termed D18); McCulloch et al. [74] (M17); Holcomb et al. [77] (H16) resulted in similar reconstructed $[CO_3^{2-}]_{CF}$ among biological replicates and that the outcome was within error of independent measurements from $[CO_3^{2-}]$ microelectrodes as reported by Sevilgen et al. [35] (S19) for the same coral species (*S. pistillata*) (M17: 909 ± 313 (SD, n = 8), H16: 757 ± 342 (SD, n = 8), D18: 848 ± 392 (SD, n = 7), S19: 679 ± 183 (SD); at 311 ppm pCO_2, 28 °C) [7].

Regression analysis of net calcification, calculated pH_{CF}, $[CO_3^{2-}]_{CF}$, and DIC_{CF} as a function of pH_{SW} were explored using an AIC approach (Figures 3 and 4; Table S9). pH_{CF} as a function of pH_{SW} follows a similar pattern to the originating $\delta^{11}B$ data (Figure 3; Table S9). Patterns in $[CO_3^{2-}]_{CF}$ as a function of pH_{SW} differ from patterns in pH_{CF} as a function of pH_{SW}, with only *P. strigosa* and *S. siderea* showing statistically significant regressions under the 28 °C treatment (Figure 4; Table S9).

In addition, regression analyses of net calcification as a function of pH_{CF}, $[CO_3^{2-}]_{CF}$, and $[DIC]_{CF}$ were explored using the AIC approach. Here, linear regressions were typically favored but rarely reached statistical significance (Table S9). This result is not surprising given the previous findings of Liu et al. [69] that pH_{CF} is not necessarily tightly coupled to net calcification responses to seawater acidification in a wide range of marine calcifiers, including the temperate coral *Oculina arbuscula*. Net calcification represents a balance of dissolution and calcification and will be influenced by both physicochemical and other controls over calcification, all of which may confound a clear relationship with pH_{CF}.

Two-way ANOVA tests were again used to explore the individual and combined effects of pCO_2 and temperature on calculated calcifying fluid parameters of pH_{CF}, $[CO_3^{2-}]_{CF}$, and DIC_{CF}. In *P. astreoides*, pCO_2 had a significant effect on pH_{CF} and $[CO_3^{2-}]_{CF}$. Temperature and the combined effect of temperature and pCO_2 had significant effects on DIC_{CF}. Similarly, in *P. strigosa*, pCO_2 had a significant effect on pH_{CF} and $[CO_3^{2-}]_{CF}$, but temperature only had an individual effect on DIC_{CF}. In *S. siderea*, the calculated pH_{CF} data did not meet assumptions of normality after transformation and, consequently, ANOVA results are not presented. There was a significant effect of pCO_2 on $[CO_3^{2-}]_{CF}$ of *S. siderea* and significant individual effects of both pCO_2 and temperature on DIC_{CF}. Lastly, U. tenuifolia displayed no significant effects of either pCO_2 or temperature on the calculated calcifying fluid parameters. Results of these analyses are presented in Table 4.

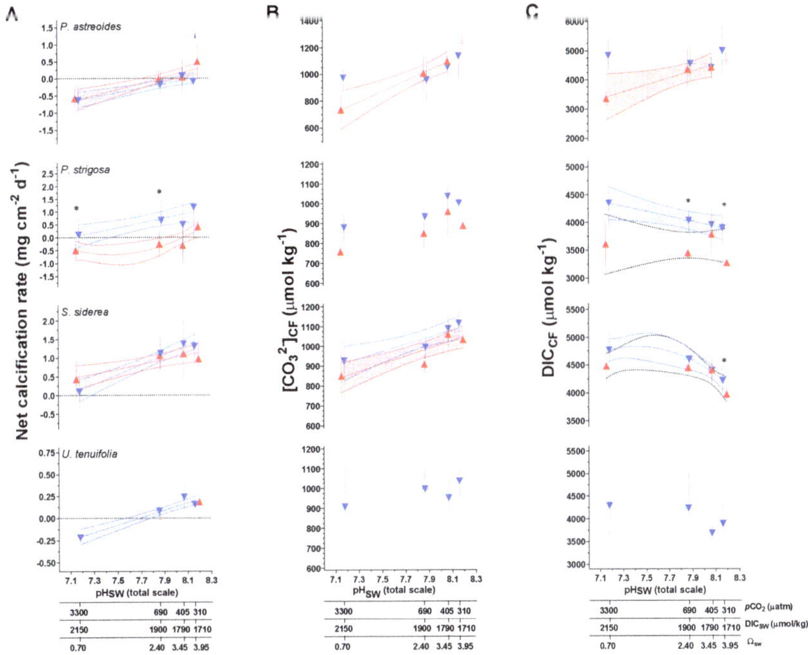

Figure 4. (**A**) Measured net calcification rate (mg cm^{-2} d^{-1}) replotted from Bove et al. [38] (**B**) B/Ca-derived carbonate ion concentration of the coral calcifying fluid ($[CO_3^{2-}]_{CF}$, μmol/mol), and (**C**) combined δ^{11}B and B/Ca-derived dissolved inorganic carbon concentration of the coral calcifying fluid (DIC_{CF}, μmol/mol). Triangle symbols (±1 SD) represent the mean value for each treatment condition, blue and red symbols represent the 28 °C and 31 °C treatment conditions, respectively. Linear versus centered quadratic fit was determined using the Akaike Information Criteria test (Table S8), with shading representing the 95% confidence interval. * indicates a significant difference between 28 °C and 31 °C data for a given pCO_2 treatment, as demonstrated by a p-value < 0.05 in a Welch's-T test.

Table 3. Net calcification data and calculated calcifying fluid pH and carbonate system parameters.

Species	T (°C)	CO_2 (μatm)	Net Calcification (mg cm^{-2} d^{-1})	Error (SD)	N [1]	pH_{CF}	Error (SD)	DIC_{CF} (mmol/mol)	Error (SD)	$[CO_3^{2-}]_{CF}$ (mmol/mol)	Error (SD)	Ω_{CF} [2]	Error (SD)	N
P. astreoides														
	28	311	−0.08	0.26	11	8.42	0.06	4983	863	1131	181	20.9	3.4	6
	28	405	0.07	0.52	12	8.45	0.06	4421	352	1033	81	19.3	1.3	6
	28	702	−0.17	0.4	10	8.39	0.10	4561	618	957	180	17.8	3.3	4
	28	3309	−0.65	0.5	12	8.35	0.05	4841	512	969	102	17.9	1.9	6
	31	288	0.5	0.5	6	8.45	0	4777	0	1242	0	23	0	1
	31	442	0.04	0.29	8	8.40	0.09	4434	394	1055	107	19.5	2	5
	31	674	−0.02	0.24	9	8.38	0.02	4365	703	1004	133	18.6	2.5	5
	31	3285	−0.54	0.27	3	8.35	0.03	3356	286	736	106	13.6	2	2
P. strigosa														
	28	312	0.92	1.28	13	8.49	0.04	3862	362	989	59	18.3	1.1	6
	28	406	1	1.12	6	8.50	0.10	3881	590	1013	85	18.7	1.6	4
	28	702	0.66	0.79	14	8.43	0.04	4025	311	933	47	17.3	0.9	6
	28	3320	0.03	0.71	16	8.36	0.08	4339	452	876	68	16.2	1.3	6
	31	288	0.41	0.3	9	8.43	0.06	3269	9	889	12	16.4	0.2	2
	31	443	−0.32	0.74	6	8.43	0.05	3782	399	959	134	17.8	2.5	5
	31	674	−0.23	0.55	7	8.42	0.03	3450	109	850	78	15.7	1.4	3
	31	3284	−0.5	0.39	8	8.37	0.10	3609	423	759	35	14	0.6	2

Table 3. *Cont.*

Species	T (°C)	CO_2 (µatm)	Net Calcification (mg cm^{-2} d^{-1})	Error (SD)	N [1]	pH_{CF}	Error (SD)	DIC_{CF} (mmol/mol)	Error (SD)	$[CO_3^{2-}]_{CF}$ (mmol/mol)	Error (SD)	Ω_{CF} [2]	Error (SD)	N
S. siderea														
	28	312	1.30	0.57	10	8.51	0.05	4216	261	1115	76	20.6	1.4	6
	28	405	1.37	0.70	12	8.47	0.03	4391	192	1087	63	20.1	1.2	6
	28	703	1.08	0.45	11	8.39	0.06	4596	194	993	129	18.4	2.4	6
	28	3317	0.12	0.52	11	8.33	0.00	4765	333	920	133	17.1	2.5	6
	31	288	1.00	0.54	8	8.46	0.03	3969	117	1033	34	19.1	0.6	4
	31	449	1.16	0.70	11	8.41	0.05	4407	252	1060	62	19.6	1.1	6
	31	673	1.06	0.51	11	8.32	0.05	4449	120	909	47	16.8	0.9	3
	31	3285	0.43	0.53	12	8.26	0.10	4480	284	851	119	15.7	2.2	5
U. tenuifolia														
	28	312	0.16	0.10	11	8.52	0.04	3886	449	1037	85	19.2	1.6	4
	28	404	0.24	0.17	7	8.49	0.04	3676	188	950	102	17.6	1.9	4
	28	698	0.05	0.08	4	8.44	0.08	4230	767	996	99	18.5	1.8	4
	28	3303	−0.23	0.05	5	8.38	0.13	4286	657	908	198	16.8	3.7	4

[1] Refers to number of specimens analyzed; [2] Ω_{CF} values calculated using the assumption of $[Ca]_{CF} = [Ca]_{SW}$.

Table 4. 2-way ANOVA output testing the influence of temperature and pCO_2 on calculated calcifying fluid carbonate system parameters and coral and symbiont physiological parameters.

Parameter/Species/Experimental Variable		*p* Value			*p* Value
Symbiont cell density			pH_{CF}		
P. astreoides			*P. astreoides*		
	Temperature	0.417		Temperature	0.676
	CO_2	0.11		CO_2	**6.54 × 10^{-5}**
	Temperature + CO_2	0.466		Temperature + CO_2	0.286
P. strigosa			*P. strigosa*		
	Temperature	**2.59 × 10^{-11}**		Temperature	0.42275
	CO_2	0.648		CO_2	**0.00536**
	Temperature: CO_2	0.15		Temperature + CO_2	0.62005
S. siderea			*S. siderea*		
	Temperature	**9.09 × 10^{-5}**		Temperature	Not Normal *
	CO_2	**0.000251**		CO_2	Not Normal *
	Temperature + CO_2	0.858348		Temperature + CO_2	Not Normal *
U. tenuifolia			*U. tenuifolia*		
	CO_2	0.727		CO_2	0.317
Chl a			$[CO_3^{2-}]$		
P. astreoides			*P. astreoides*		
	Temperature	**9.89 × 10^{-7}**		Temperature	0.41309
	CO_2	**6.10 × 10^{-11}**		CO_2	**0.00333**
	Temperature + CO_2	**2.80 × 10^{-9}**		Temperature + CO_2	0.22932
P. strigosa			*P. strigosa*		
	Temperature	**6.15 × 10^{-8}**		Temperature	0.0988
	CO_2	**0.000217**		CO_2	**0.0224**
	Temperature: CO_2	0.070441		Temperature: CO_2	0.855
S. siderea			*S. siderea*		
	Temperature	**0.00394**		Temperature	0.679
	CO_2	**1.37 × 10^{-7}**		CO_2	**5.98 × 10^{-5}**
	Temperature + CO_2	0.25756		Temperature + CO_2	0.996
U. tenuifolia			*U. tenuifolia*		
	CO_2	**0.033**		CO_2	0.181
Total host energy			$[DIC]_{CF}$		
P. astreoides			*P. astreoides*		
	Temperature	**0.02755**		Temperature	**0.00183**
	CO_2	**0.00659**		CO_2	0.69095
	Temperature: CO_2	0.06197		Temperature + CO_2	**0.04313**
P. strigosa			*P. strigosa*		
	Temperature	**8.35 × 10^{-5}**		Temperature	**0.00395**
	CO_2	0.103		CO_2	0.26074
	Temperature + CO_2	0.857		Temperature + CO_2	0.3616

Table 4. *Cont.*

Parameter/Species/Experimental Variable		p Value			p Value
S. siderea			*S. siderea*		
	Temperature	**0.000522**		Temperature	**0.00119**
	CO_2	0.118746		CO_2	$\mathbf{2.62 \times 10^{-5}}$
	Temperature + CO_2	0.710086		Temperature + CO_2	0.3765
U. tenuifolia			*U. tenuifolia*		
	CO_2	0.168		CO_2	0.463
Calcification rate					
P. astreoides					
	Temperature	**0.0178**			
	CO_2	$\mathbf{5.11 \times 10^{-6}}$			
	Temperature + CO_2	0.1779			
P. strigosa					
	Temperature	**0.000142**			
	CO_2	**0.000364**			
	Temperature + CO_2	0.678234			
S. siderea					
	Temperature	0.52			
	CO_2	$\mathbf{3.27 \times 10^{-8}}$			
	Temperature + CO_2	0.209			

* Data did not reach normality even after transformation steps, so ANOVA test was not performed.

We combined our data with a compilation of coral data from the literature using different approaches (geochemistry, microelectrode, pH-sensitive dye; Table S11). It is clear from almost all published studies that pH_{CF}, $[CO_3^{2-}]_{CF}$, and DIC_{CF} are differentiated from seawater values and that this is an active compensatory process as pH_{CF}, and $[CO_3^{2-}]_{CF}$ values are increasingly elevated relative to seawater values under acidified conditions (Figures 5 and 6).

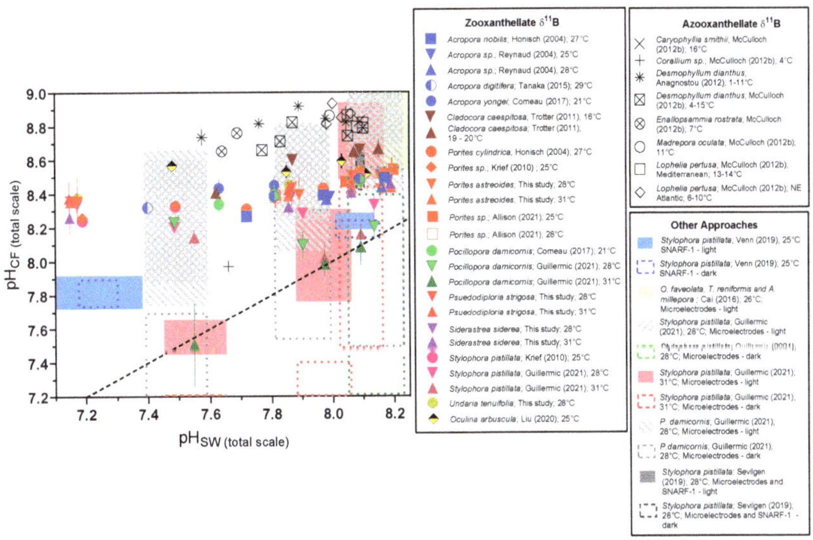

Figure 5. Compilation of δ^{11}B-derived calcification fluid pH_{CF} data from shallow-water zooxanthellate coral culturing experiments. We include data from cold-water azooxanthellate corals as a point of reference, as well as ranges of pH_{CF} data as determined by SNARF pH-sensitive dyes and pH microelectrode analyses. 'RR' label denotes corals treated with the Ca-ATPase inhibitor ruthenium red [32]. Previously published data used in the figure are available in Table S12 alongside citations for data sources.

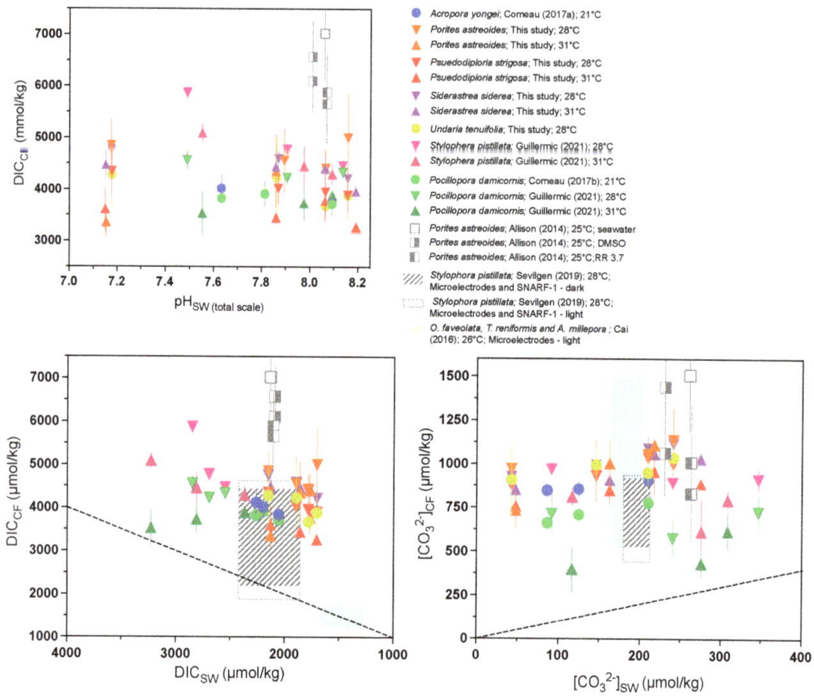

Figure 6. Compilation of combined $\delta^{11}B$ and B/Ca-derived DIC_{CF} and $[CO_3{}^{2-}]_{CF}$ measurements from available published data, as well as published microelectrode data. We include cultured shallow-water zooxanthellate corals, cold-water azooxanthellate corals, as well as a range of DIC_{CF} values as determined by SNARF-1 pH-sensitive dyes and pH microelectrode analyses. (**A**) DIC_{CF} as a function of pH_{SW} (total scale), (**B**) DIC_{CF} as a function of DIC_{SW}, and (**C**) $[CO_3{}^{2-}]_{CF}$ as a function of $[CO_3{}^{2-}]_{SW}$. 'RR' label denotes corals treated with the Ca-ATPase inhibitor ruthenium red [32]. Previously published data used in the figure are available in Table S12 alongside citations for data sources.

3.3. Biological Compensation for External pH Changes, and the Importance of Coral and Symbiont Physiology

Changes in pH_{CF} across experimental conditions are relatively small, with values changing, for example, from 8.52 to 8.26 in the most sensitive coral species *S. siderea*. However, it is important to note that this represents a substantial compensation or buffering by the coral with respect to the much larger changes in pH_{SW}, which range from 8.3 to 7.3 in the experiment.

To illustrate the chemical effects of this compensation, we provide calculations of the proton differential ($[H^+]_{SW} - [H^+]_{CF}$), maintained by the coral with respect to seawater under all experimental conditions (Figure 7, Table S9), for which larger values represent a greater difference in proton concentration in the calcification fluid relative to seawater. For example, for *P. astreoides* at 28 °C, the comparison between the 311 μatm pCO_2 and the ~33,309 μatm pCO_2 treatments show a proton differential of 0.5×10^{-8} mol/L at 400 μatm pCO_2 and 6.3×10^{-8} mol/L at 3000 μatm, which represents a differential 12.4 times higher at the elevated pCO_2 condition. At 31 °C for *P. astreoides*, the differential is 13.4 times higher between the ~400 μatm and ~3000 μatm pCO_2 conditions. The same comparison shows a proton differential 11.4 and 13.4 times higher for *P. strigosa* at 28 °C and 31 °C, 11.9 and 13.9 times higher for *S. siderea* at 28 °C and 31 °C and 11.5 times higher for *U. tenuifolia* at 28 °C.

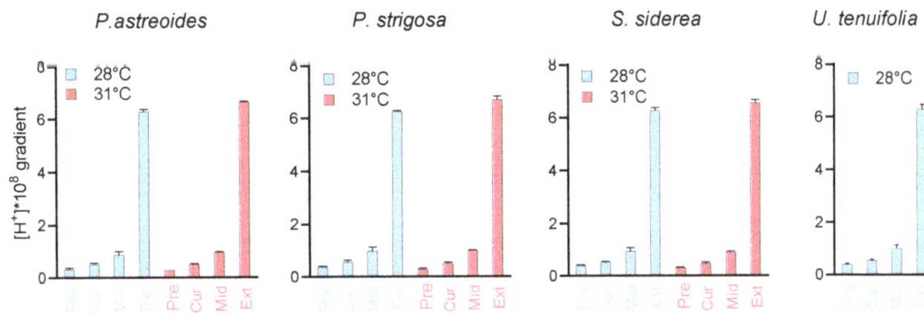

Figure 7. Proton differentials maintained by the coral species between seawater and their calcifying fluid ($[H^+]_{SW} - [H^+]_{CF}$). Values are given for preindustrial, present day, mid-century, and extreme (~3300 ppm) pCO_2 conditions, with precise pCO_2 values given in Table S1. Calculated values are provided in Table S9.

Using the compilation of pH_{CF} data from the literature and from the present study, we compare the slopes of the relationships between pH_{CF} and pH_{SW} (Figure 8), with a slope of 1 indicating that changes in pH_{CF} perfectly track changes in pH_{SW}, and a slope of 0 indicating that the coral is perfectly compensating for external changes in pH_{SW} and that pH_{CF} is invariant across the experimental conditions. This compilation demonstrates that although there are cases where corals perfectly compensate for external pH_{SW} changes, in most cases corals exhibit a modest decrease in pH_{CF} with decreasing pH_{SW} (Figure 8). Furthermore, $\delta^{11}B$-based estimates of pH_{CF} were generally lower than that found by the microelectrode approach (Figure 8). We hypothesized in previous work that this could reflect different time intervals being recorded by these approaches [7], but it cannot be ruled out that there are systematic differences in the techniques, such as recording different microenvironments.

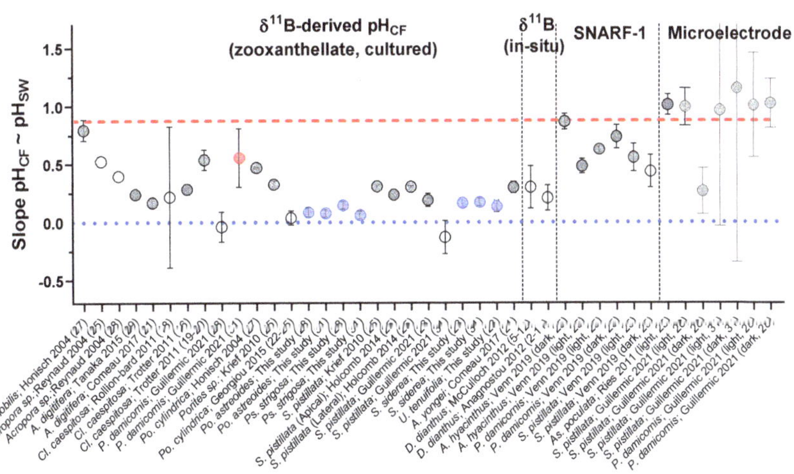

Figure 8. The sensitivity of pH_{CF} to pH_{SW} within the coral species studied here and reported in the literature. Y-Axis is the slope of the relationship between pH_{CF} and pH_{SW}, with a value of 1 showing that change in pH_{CF} directly tracks the magnitude of the change in pH_{SW}, and a value of 0 indicating perfect control over coral pH_{CF} amidst variable pH_{SW}. On X axis, numbers in parenthesis are the temperature of the experiment. Previously published data used in the figure are available in Table S12 alongside citations for data sources.

It should ultimately be Ω_{CF} that is the most influential factor in coral calcification. If pH_{CF} modestly decreases, but Ω_{CF} remains high enough to support efficient coral calcification, then this may explain how corals continue calcifying despite acidification of their calcifying fluid. Calculating Ω_{CF} requires an additional assumption due to the lack of constraints on $[Ca^{2+}]_{CF}$. We conducted a set of calculations where $[Ca^{2+}]_{CF}$ is assumed to be equivalent to $[Ca^{2+}]_{SW}$ (Table 2 and Table S7). The calculated saturation states were then plotted against coral calcification data (Figure 9). Our compilation of Ω_{CF} exhibits inter-species variation, but none of the four species (*S. siderea*, *P. strigosa*, *P. astreoides* and *U. tenuifolia*) exhibit significantly different values from each other at under control temperature and pCO_2 conditions (ANOVA test, $p = 0.2$), but were significantly higher compared to that calculated for *S. pistillata* [7], A. yongei ($p < 0.05$) and *P. damicornis* ($p < 0.05$) [7,67] using the same assumptions. Species-specific threshold values of Ω_{CF}, where calcification is outweighed by dissolution, can be estimated from Figure 9, but care should be taken interpreting those absolute values due to the lack of constraints on $[Ca^{2+}]_{CF}$. There is a possibility that $[Ca^{2+}]_{CF}$ could differ between species, as highlighted by previous work on A. yongei and *P. damicornis* [51]. In addition, the highest calculated Ω_{CF} values do not translate to highest calcification rates for the corals, which may reflect prior observations that coral calcification is complex process beyond simple physicochemical manipulation of the calcifying fluid [78,82].

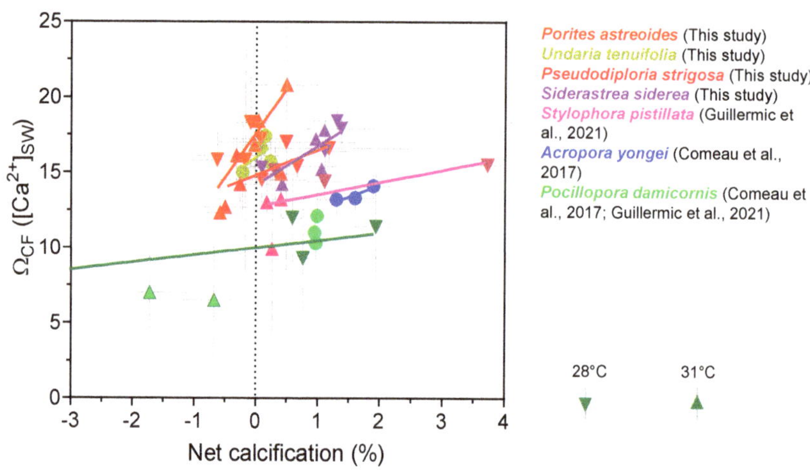

Figure 9. Relationships between boron isotope derived Ω_{CF} and net calcification rates. Ω_{CF} is calculated assuming that $[Ca^{2+}]$ of seawater is equivalent to $[Ca^{2+}]$ of the calcifying fluid. Published data used in this figure is given in Table S12 with citations for data sources.

Statistical analyses of coral and symbiont physiological data from this experiment have been explored elsewhere [39]. Here, we perform additional 2-way ANOVA tests on the same data to produce statistical tests that are handled identically between the geochemical and physiological data. In *P. astreoides*, we found significant individual and combined effects of pCO_2 and temperature on Chl a, and total host energy reserves, but not on symbiont cell density. Significantly, and as noted in previous work of Bove et al. [39], the significant effect of temperature on Chl a in *P. astreoides* is actually to increase levels, whereas for all other significant effects reported from our ANOVA test they are characterizing decreases. However, it is important to note that temperature actually improved symbiont physiological parameters in this species, rather than repressing them as reported previously for this species [39]. In *P. strigosa*, only temperature had an effect on both symbiont cell density and total host energy reserves. Significant individual effects of both temperature and pCO_2 on both symbiont cell density and Chl a were observed for *S. siderea*, whilst only temperature had an effect on total host energy for this species. For *U. tenuifolia*, only pCO_2

had an effect on Chl a. Results of these analyses are presented in Table 4. Specimens of *S. siderea* were collected from both the inshore and offshore environments and previous work showed that reef environment affected physiological response; however, our results show a nonsignificant effect of reef environment on physiological data (Table S10), which likely reflects the difference in statistical approaches, with ANOVA tests employed here on single parameters versus the principal component analyses conducted by Bove et al. [39].

These physiological data are relevant for understanding how host and symbionts responses to environmental change may influence the chemistry of the calcifying fluid. It is notable that in the three coral species examined (excluding *U. tenuifolia*), temperature was found to influence total host energy reserves, or the total of the lipid, carbohydrate, and protein concentrations for a given area of coral host tissue. Temperature was not found to influence pH_{CF} and $[CO_3{}^{2-}]_{CF}$, but in all three corals was found to reduce $[DIC]_{CF}$ (Table 4). Whilst this observation does not prove cause and effect, it is consistent with temperature stress causing depletion of host energy reserves, and reduced host input of respiration produced CO_2 into the DIC pool for calcification. As temperature increase was associated with reduced symbiont cell density and Chl a in some of the corals, this is consistent with a scenario where the symbiont was less efficiently supplying energy to the host, leading to the observed decrease in energy reserves.

To further explore the relationship between these host and symbiont physiological changes and the source of carbon in the calcification fluid, we examined $\Delta\delta^{13}C$ ($\delta^{13}C_{31^\circ C}$ − $\delta^{13}C_{28^\circ C}$) for each pCO_2 treatment. In Figure 10, $\Delta\delta^{13}C$ is plotted against Δ[DIC] ($[DIC]_{31^\circ C}$ − $[DIC]_{28^\circ C}$). This analysis shows that, on one hand, there is a significant difference in the source of carbon for calcification between *S. siderea* and *S. pistillata*, for which temperature did not influence $\delta^{13}C$ at the higher pCO_2 conditions. On the other hand, it shows that at higher pCO_2 conditions, $\Delta\delta^{13}C$ of *P. astreoides* and *P. strigosa* (Figure 10) shifted to more positive values and that Δ[DIC] shifted towards more negative values compared to the other corals, with the overall effect most pronounced in *P. astreoides*. The potential mechanisms driving these trends are discussed below.

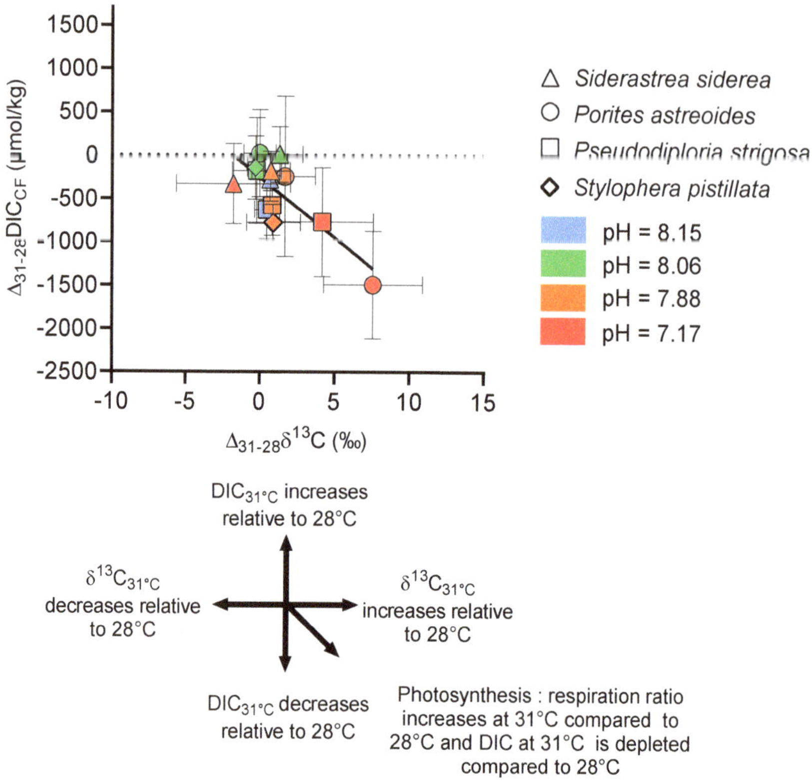

Figure 10. Combined influence of temperature and pCO_2 on the $\delta^{13}C$ and [DIC] of the coral calcifying fluid. Data are expressed as $\Delta\delta^{13}C$ ($\delta^{13}C_{31°C} - \Delta\delta^{13}C_{28°C}$) and Δ[DIC] ($[DIC]_{31°C} - [DIC]_{28°C}$) for each pCO_2 treatment.

4. Discussion

Collectively, the data from this study and the literature suggest that most corals experience a degree of acidification of their internal calcification fluid media in response to external seawater acidification. Nonetheless, considerable biological compensation is observed for the studied species, which maintain a larger proton differential between the calcifying fluid and seawater under elevated pCO_2. Therefore, the slight decreases in pH_{CF} under elevated pCO_2 may represent either a practical tradeoff by the coral, or a biological limit to controlling calcifying fluid chemistry that still allows the coral to maintain $[CO_3^{2-}]_{CF}$ and Ω_{CF} at a level that supports efficient calcification. The hypothesis that this is an active and energy-consuming adaptation by corals is supported by the finding that the ability of corals to regulate pH and carbonate chemistry of their calcifying fluid can be impaired by thermal stress and symbiont loss [7,25]. In contrast to our studies of *S. pistillata* and *P. damicornis* [7,25], we found that temperature did not result in a statistically significant change in pH_{CF} regulation across all conditions in the Caribbean species investigated here, although it did under a subset of the experimental conditions in *S. siderea*. This may reflect that the highest temperature treatment in the present experiment [38,39], did not elicit as severe an effect on the corals and their symbionts as in the prior experiment [7,25], with the exception of *U. tenuifolia* that had too high a mortality at elevated temperature to be examined. Notably, the Ca-ATPase inhibitor experiments of Allison et al. [32], and the heat stress experiments of Guillermic et al. [7], on *Pocillopora damicornis* are the only experiments where corals appeared to almost completely lose the ability to elevate pH_{CF} relative to seawater, as shown in Figure 5.

Furthermore, we are able to examine associations between coral and symbiont physiological responses and calcifying fluid regulation. This revealed a significant depletion in total host energy reserves at elevated temperature, as well as a reduction in $[DIC]_{CF}$ in all three corals examined. The reduction in calculated $[DIC]_{CF}$ with temperature increase was also seen in the *S. pistillata* and *P. damicornis* experiment described in Guillermic et al. [7]. This result is consistent with a reduction in the energy transfer between the symbiont and host leading to the host depleting its energy reserves and/or the host consuming more energy in response to temperature stress. Although we cannot prove cause and effect, changes in the energy balance between host and symbiont could then lead to a reduction in the production of metabolic CO_2 that is contributed to the DIC pool for calcification, which is consistent with the reduction we see in calculated $[DIC]_{CF}$ in these corals. The story is also more complex, however, with additional interactions apparent from the $\delta^{13}C$ data.

Although the $\delta^{13}C$ data did not indicate major differences in host and symbiont physiological influence on $[DIC]_{CF}$ for the different coral species under the control CO_2 conditions, significant differences in $\delta^{13}C$ emerged under elevated temperature and CO_2 amongst the species, with shifts towards positive $\delta^{13}C$ associated with amplified reductions in $[DIC]_{CF}$ in two of the coral species studied.

It is thought that changes in the flux of carbon to and from photosynthesis and respiration, respectively, may change the $\delta^{13}C$ of the DIC pool used for calcification. Photosynthesis preferentially consumes ^{12}C, which may then result in ^{13}C enrichment in the DIC pool for calcification if this pool acts as a source of CO_2 for photosynthesis [83]. As the rate of photosynthesis increases and/or the rate of respiration decreases, the DIC pool should have more positive $\delta^{13}C$ resulting in more positive coral skeletal $\delta^{13}C$ [83,84]. In the present experiment, we can see that host respiration input is likely decreasing as the host consumes energy reserves at elevated temperature. There is also an overall reduction in $[DIC]_{CF}$, which is likely due to a reduction in CO_2-input from respiration. One way to explain the relationship between the $\Delta\delta^{13}C$ and $\Delta[DIC]$ data in Figure 10 is that, in addition to the reduction in respiration-sourced CO_2 to the DIC pool for calcification, photosynthetic drawdown of $^{12}CO_2$ by the symbiont is significantly enriching the $\delta^{13}C$ of the DIC pool in *P. astreoides* and *P. strigosa* at highest temperature and pCO_2 conditions, which could further deplete $[DIC]_{CF}$ and drive more positive $\Delta\delta^{13}C$. This effect was not seen in a similar experiment on *S. pistillata* [7] where symbiont loss and bleaching of the coral occurred [25], which suggests that photosynthetic influence over the DIC pool was minimal at the higher temperature and pCO_2 conditions. This interpretation is also congruent with observations from this experiment that the symbiont in *P. astreoides* is not negatively impacted by increased temperature [39] and is evidence for a divergent host and symbiont physiological response to environmental parameters in this coral species. Conversely, the effects of host respiration and symbiont photosynthesis remained relatively balanced in terms of their influence on the source of DIC for calcification in *S. siderea*. It has been previously noted that *P. astreoides*, *S. siderea*, and *P. strigosa* host differing algal symbiont communities which may result in differing carbon allocation between the host and symbiont [39]. For example, *S. siderea* mainly hosts *Cladocopium* and *Breviolum* whereas *P. astreoides* hosts *Breviolum* and *Symbiodinium* [39]. The potential changes in carbon flux between the host and symbiont is interesting to consider in the context of the debate on the extent of mutualism between the coral host and symbiont, and whether the symbiont develops features that are more parasitic in nature under environmental change [39,85,86].

Overall, the results presented here and in previous work on these corals [7,38,39,87] highlight the considerable complexity in host and symbiont responses to ocean warming and acidification, demonstrating that these responses are closely coupled to a coral's physicochemical control of their calcification process. The combined physiological data from this study, preceding studies [38,39,87], as well as the work of other groups (e.g., [88]), converge on the conclusion that *S. siderea* is a relatively resilient species to environmental change and so may be amongst the least impacted by future warming and acidification, compared to a number of more sensitive Caribbean species studied such as *O. faveolata*, *M. cavernosa*, and *U tenuifolia* [38,88]. Some differences in observations on *P. astreoides* between studies exist in its temperature sensitivity [38,88] which may reflect intraspecific differences in host or symbiont physiology, with the evidence here suggesting thermal sensitivity can be relatively low for the symbiont of at least some Caribbean *P. astreoides* but that reductions in net calcification can still occur in non-bleached corals in acidified seawater.

The type of data presented here provides useful insights into the underlying physiological mechanisms of coral adaptation, and the variation in responses between species in an experiment. Whilst difficult to make regional or global predictions from single studies, a recent metanalyses of experimental data suggested that at the levels predicted over the next century acidification is less of a threat to Caribbean corals compared to warming but that there is evidence for variation across the region and considerable data gaps [89]. Despite this finding, the evidence presented in this paper suggest acidification resilient Caribbean corals maintain increasingly large proton differentials with seawater under acidification, highlighting that there may be an energetic cost to this adaption that corals would have to maintain indefinitely in the future. The work here also highlights that warming can have complex influences on the calcification processes even outside a severe bleaching event through changes to the partitioning of carbon between the host, symbiont, and the pool of ions used for calcification. The sustainability of these adaptations to acidification and warming are largely unexplored in Caribbean corals, but is highly relevant given findings from this experiment that even if positive net calcification is maintained in environmental challenge experiments corals could be depleting energy reserves [39]. Therefore, it will be important for future research to attempt to determine whether even the relatively resilient species such as *S. siderea* can maintain their short-term adaptive responses over the longer term.

Overall, the results presented here and in previous work on these corals [7,38,39,87] highlight the considerable complexity in host and symbiont responses to ocean warming and acidification, and demonstrate that these responses are closely coupled to a coral's physicochemical control of their calcification process.

Supplementary Materials: The following supporting information can be downloaded at: https://www.mdpi.com/article/10.3390/jmse10081075/s1, Table S1: Measured Culture Conditions in the 8 Experimental Setups from Bove et al., 2019; Table S2: Calculated Culture Conditions in the 8 Experimental Setups from Bove et al., 2019; Table S3: Physiological data of the host and the symbionts originally published in Bove et al. (2021) from the same corals used in this study; Table S4: Boron isotope standard reproducibility; Table S5: Reproducibility of B/Ca analyses of CamWuellestorfi and NEP standards; Table S6: Functions used for data transformations according the method of Tukey et al., 1977; Table S7: Geochemical data and calculated parameters from individual Caribbean coral specimens, of which the average per experimental condition are given in main text Tables 1 and 2; Table S8: Average and individual coral specimen geochemical data and calculations from *Stylophera pistillata*; Table S9: AIC results for Caribbean coral geochemical data as a function of the pH of seawater in the experiment and net calcification as a function of calculated CF parameter data; Table S10: Calculations of pH gradient between seawater and the calcifying fluid; Table S11: ANOVA results for Caribbean coral geochemical and physiological data; Table S12: Compilation of published data used in the main text.

Author Contributions: Conceptualization, R.A.E., K.D.C. and J.B.R.; methodology, R.A.E., S.M., K.D.C. and J.B.R.; formal analysis, C.B.B., L.P.C., S.M., B.A.C., M.G., I.D.C. and R.A.E.; investigation, C.B.B., L.P.C., M.G., I.D.C., K.D.C., R.A.E. and J.B.R.; resources, R.A.E., K.D.C. and J.B.R.; data curation, M.G. and R.A.E.; writing—original draft preparation, R.A.E. and I.D.C.; writing—review and editing, all authors; visualization, M.G., I.D.C. and R.A.E.; supervision, R.A.E., K.D.C. and J.B.R.; project administration, R.A.E., K.D.C. and J.B.R.; funding acquisition, R.A.E., K.D.C. and J.B.R. All authors have read and agreed to the published version of the manuscript.

Funding: RAE and JBR acknowledge support from National Science Foundation grants OCE-1437166 and OCE-1437371. KDC acknowledges support from NSF grant OCE-1459522. RAE, MG, ID, and BA were also supported by the "Laboratoire d'Excellence" LabexMER (ANR-10-LABX-19) Chair to RAE, co-funded by a grant from the French government under the program "Investissements d'Avenir" and MG was supported by an IAGC student grant 2017. RAE acknowledges support from the Pritzker Endowment to UCLA IoES, and JBR acknowledges support from the ZMT and the Hanse-Wissenschaftskolleg Fellowship Program. BA was supported by an NSF GRFP. M.G., I.D., and BA were supported by the Center for Diverse Leadership in Science, funded by NSF, the Packard Foundation, Sloan Foundation, and Silicon Valley Community Foundation.

Data Availability Statement: Data will be archived in the BCO-DMO archive on publication of this article (https://www.bco-dmo.org/, accessed on 24 July 2022).

Acknowledgments: We thank Belize Fisheries Department for all associated permits, the Toledo Institute for Development and Environment (TIDE), and the Southern Environmental Association (SEA) for their continued support, as well as Garbutt's Marine for assistance in the field.

Conflicts of Interest: The authors declare no conflict of interest. The funders had no role in the design of the study; in the collection, analyses, or interpretation of data; in the writing of the manuscript, or in the decision to publish the results.

References

1. Glynn, P.W. Coral reef bleaching in the 1980s and possible connections with global warming. *Trends Ecol. Evol.* **1991**, *6*, 175–179. [CrossRef]
2. Hughes, T.P.; Kerry, J.T.; Álvarez-Noriega, M.; Álvarez-Romero, J.G.; Anderson, K.D.; Baird, A.H.; Babcock, R.C.; Beger, M.; Bellwood, D.R.; Berkelmans, R.; et al. Global warming and recurrent mass bleaching of corals. *Nature* **2017**, *543*, 373–377. [CrossRef] [PubMed]
3. Doney, S.C.; Fabry, V.J.; Feely, R.A.; Kleypas, J.A. Ocean Acidification: The Other CO_2 Problem. *Annu. Rev. Mar. Sci.* **2009**, *1*, 169–192. [CrossRef] [PubMed]
4. Iglesias-Rodriguez, M.D.; Halloran, P.R.; Rickaby, R.E.M.; Hall, I.R.; Colmenero-Hidalgo, E.; Gittins, J.R.; Green, D.R.H.; Tyrrell, T.; Gibbs, S.J.; von Dassow, P.; et al. Phytoplankton Calcification in a High-CO_2 World. *Science* **2008**, *320*, 336–340. [CrossRef]
5. Ries, J.B.; Cohen, A.L.; McCorkle, D.C. Marine calcifiers exhibit mixed responses to CO_2-induced ocean acidification. *Geology* **2009**, *37*, 1131–1134. [CrossRef]
6. Kroeker, K.; Micheli, F.; Gambi, M.C. Ocean acidification causes ecosystem shifts via altered competitive interactions. *Nat. Clim. Chang.* **2012**, *3*, 156–159. [CrossRef]
7. Guillermic, M.; Cameron, L.P.; De Corte, I.; Misra, S.; Bijma, J.; de Beer, D.; Reymond, C.E.; Westphal, H.; Ries, J.B.; Eagle, R.A. Thermal stress reduces pocilloporid coral resilience to ocean acidification by impairing control over calcifying fluid chemistry. *Sci. Adv.* **2021**, *7*, eaba9958. [CrossRef]
8. Marubini, F.; Atkinson, M. Effects of lowered pH and elevated nitrate on coral calcification. *Mar. Ecol. Prog. Ser.* **1999**, *188*, 117–121. [CrossRef]
9. Langdon, C.; Takahashi, T.; Sweeney, C.; Chipman, D.; Goddard, J.; Marubini, F.; Aceves, H.; Barnett, H.; Atkinson, M.J. Effect of calcium carbonate saturation state on the calcification rate of an experimental coral reef. *Glob. Biogeochem. Cycles* **2000**, *14*, 639–654. [CrossRef]
10. Langdon, C. Effect of elevated pCO_2 on photosynthesis and calcification of corals and interactions with seasonal change in temperature/irradiance and nutrient enrichment. *J. Geophys. Res. Earth Surf.* **2005**, *110*, C9. [CrossRef]
11. Marubini, F.; Barnett, H.; Langdon, C.; Atkinson, M.J. Dependence of calcification on light and carbonate ion concentration for the hermatypic coral Porites compressa. *Mar. Ecol. Prog. Ser.* **2001**, *220*, 153–162. [CrossRef]
12. Renegar, D.; Riegl, B. Effect of nutrient enrichment and elevated CO_2 partial pressure on growth rate of Atlantic scleractinian coral Acropora cervicornis. *Mar. Ecol. Prog. Ser.* **2005**, *293*, 69–76. [CrossRef]
13. Schneider, K.; Erez, J. The effect of carbonate chemistry on calcification and photosynthesis in the hermatypic coral Acropora eurystoma. *Limnol. Oceanogr.* **2006**, *51*, 1284–1293. [CrossRef]
14. Anthony, K.R.N.; Kline, D.I.; Diaz-Pulido, G.; Dove, S.; Hoegh-Guldberg, O. Ocean acidification causes bleaching and productivity loss in coral reef builders. *Proc. Natl. Acad. Sci. USA* **2008**, *105*, 17442–17446. [CrossRef] [PubMed]
15. Jokiel, P.L.; Rodgers, K.S.; Kuffner, I.B.; Andersson, A.J.; Cox, E.F.; Mackenzie, F.T. Ocean acidification and calcifying reef organisms: A mesocosm investigation. *Coral Reefs* **2008**, *27*, 473–483. [CrossRef]
16. Marubini, F.; Ferrier-Pagès, C.; Furla, P.; Allemand, D. Coral calcification responds to seawater acidification: A working hypothesis towards a physiological mechanism. *Coral Reefs* **2008**, *27*, 491–499. [CrossRef]
17. Krief, S.; Hendy, E.J.; Fine, M.; Yam, R.; Meibom, A.; Foster, G.L.; Shemesh, A. Physiological and isotopic responses of scleractinian corals to ocean acidification. *Geochim. Cosmochim. Acta* **2010**, *74*, 4988–5001. [CrossRef]

18. Holcomb, M.; Cohen, A.L.; McCorkle, D.C. An investigation of the calcification response of the scleractinian coral Astrangia poculata to elevated pCO_2 and the effects of nutrients, zooxanthellae and gender. *Biogeosciences* **2012**, *9*, 29–39. [CrossRef]
19. Edmunds, P.J.; Brown, D.; Moriarty, V. Interactive effects of ocean acidification and temperature on two scleractinian corals from Moorea, French Polynesia. *Glob. Chang. Biol.* **2012**, *18*, 2173–2183. [CrossRef]
20. Comeau, S.; Carpenter, R.C.; Edmunds, P.J. Effects of irradiance on the response of the coral Acropora pulchra and the calcifying alga Hydrolithon reinboldii to temperature elevation and ocean acidification. *J. Exp. Mar. Biol. Ecol.* **2014**, *453*, 28–35. [CrossRef]
21. Reynaud, S.; Leclercq, N.; Romaine-Lioud, S.; Ferrier-Pagès, C.; Jaubert, J.; Gattuso, J.-P. Interacting effects of CO_2 partial pressure and temperature on photosynthesis and calcification in a scleractinian coral. *Glob. Chang. Biol.* **2003**, *9*, 1660–1668. [CrossRef]
22. Rodolfo-Metalpa, R.; Martin, S.; Ferrier-Pagès, C.; Gattuso, J.-P. Response of the temperate coral Cladocora caespitosa to mid- and long-term exposure to pCO_2 and temperature levels projected for the year 2100 AD. *Biogeosciences* **2010**, *7*, 289–300. [CrossRef]
23. Edmunds, P.J. Zooplanktivory ameliorates the effects of ocean acidification on the reef coral *Porites* spp. *Limnol. Oceanogr.* **2011**, *56*, 2402–2410. [CrossRef]
24. Schoepf, V.; Grottoli, A.G.; Warner, M.E.; Cai, W.-J.; Melman, T.F.; Hoadley, K.D.; Pettay, D.T.; Hu, X.; Li, Q.; Xu, H.; et al. Coral Energy Reserves and Calcification in a High-CO_2 World at Two Temperatures. *PLoS ONE* **2013**, *8*, e75049. [CrossRef] [PubMed]
25. Cameron, L.P.; Reymond, C.E.; Bijma, J.; Büscher, J.; De Beer, D.; Guillermic, M.; Eagle, R.A.; Gunnel, J.; Müller-Lundin, F.; Schmidt-Greib, G.; et al. Impacts of warming and acidification on coral calcification linked to symbiont loss and deregulation of calcifying fluid pH. *J. Mar. Sci. Eng.* **2022**, *unpublished*.
26. Innis, T.; Allen-Waller, L.; Brown, K.T.; Sparagon, W.; Carlson, C.; Kruse, E.; Huffmyer, A.S.; Nelson, C.E.; Putnam, H.M.; Barott, K.L. Marine heatwaves depress metabolic activity and impair cellular acid–base homeostasis in reef-building corals regardless of bleaching susceptibility. *Glob. Chang. Biol.* **2021**, *27*, 2728–2743. [CrossRef]
27. D'Olivo, J.P.; McCulloch, M. Response of coral calcification and calcifying fluid composition to thermally induced bleaching stress. *Sci. Rep.* **2017**, *7*, 2207. [CrossRef]
28. D'Olivo, J.P.; Ellwood, G.; DeCarlo, T.M.; McCulloch, M. Deconvolving the long-term impacts of ocean acidification and warming on coral biomineralisation. *Earth Planet. Sci. Lett.* **2019**, *526*, 115785. [CrossRef]
29. Schoepf, V.; D'Olivo, J.P.; Rigal, C.; Jung, E.M.U.; McCulloch, M.T. Heat stress differentially impacts key calcification mechanisms in reef-building corals. *Coral Reefs* **2021**, *40*, 459–471. [CrossRef]
30. Venn, A.; Tambutté, E.; Holcomb, M.; Allemand, D.; Tambutté, S. Live Tissue Imaging Shows Reef Corals Elevate pH under Their Calcifying Tissue Relative to Seawater. *PLoS ONE* **2011**, *6*, e20013. [CrossRef]
31. Venn, A.A.; Tambutté, E.; Holcomb, M.; Laurent, J.; Allemand, D.; Tambutté, S. Impact of seawater acidification on pH at the tissue–skeleton interface and calcification in reef corals. *Proc. Natl. Acad. Sci. USA* **2012**, *110*, 1634–1639. [CrossRef] [PubMed]
32. Allison, N.; Cohen, I.; Finch, A.A.; Erez, J.; Tudhope, A.W. Corals concentrate dissolved inorganic carbon to facilitate calcification. *Nat. Commun.* **2014**, *5*, 5741. [CrossRef] [PubMed]
33. McCulloch, M.T.; D'Olivo, J.P.; Falter, J.; Georgiou, L.; Holcomb, M.; Montagna, P.; Trotter, J.A. Boron Isotopic Systematics in Scleractinian Corals and the Role of pH Up-regulation. In *Boron Isotopes. Advances in Isotope Geochemistry*; Springer: Cham, Switzerland, 2017; pp. 145–162. [CrossRef]
34. Venn, A.A.; Tambutté, E.; Caminiti-Segonds, N.; Techer, N.; Allemand, D. Effects of light and darkness on pH regulation in three coral species exposed to seawater acidification. *Sci. Rep.* **2019**, *9*, 2201. [CrossRef] [PubMed]
35. Sevilgen, D.S.; Venn, A.A.; Hu, M.Y.; Tambutté, E.; de Beer, D.; Planas-Bielsa, V.; Tambutté, S. Full in vivo characterization of carbonate chemistry at the site of calcification in corals. *Sci. Adv.* **2019**, *5*, eaau7447. [CrossRef]
36. Guo, W. Seawater temperature and buffering capacity modulate coral calcifying pH. *Sci. Rep.* **2019**, *9*, 1189. [CrossRef] [PubMed]
37. Chalk, T.B.; Standish, C.D.; D'Angelo, C.; Castillo, K.D.; Milton, J.A.; Foster, G.L. Mapping coral calcification strategies from in situ boron isotope and trace element measurements of the tropical coral Siderastrea siderea. *Sci. Rep.* **2021**, *11*, 472. [CrossRef]
38. Bove, C.B.; Ries, J.B.; Davies, S.W.; Westfield, I.T.; Umbanhowar, J.; Castillo, K.D. Common Caribbean corals exhibit highly variable responses to future acidification and warming. *Proc. R. Soc. B Boil. Sci.* **2019**, *286*, 20182840. [CrossRef]
39. Bove, C.B.; Davies, S.W.; Ries, J.B.; Umbanhowar, J.; Thomasson, B.C.; Farquhar, E.B.; McCoppin, J.A.; Castillo, K.D. Global change differentially modulates coral physiology and suggests future shifts in Caribbean reef assemblages. *bioRxiv* **2021**. [CrossRef]
40. Castillo, K.D.; Lima, F.P. Comparison of in situ and satellite-derived (MODIS-Aqua/Terra) methods for assessing temperatures on coral reefs. *Limnol. Oceanogr. Methods* **2010**, *8*, 107–117. [CrossRef]
41. Barker, S.; Greaves, M.; Elderfield, H. A study of cleaning procedures used for foraminiferal Mg/Ca paleothermometry. *Geochem. Geophys. Geosyst.* **2003**, *4*. [CrossRef]
42. Gaillardet, J.; Lemarchand, D.; Göpel, C.; Manhès, G. Evaporation and Sublimation of Boric Acid: Application for Boron Purification from Organic Rich Solutions. *Geostand. Geoanal. Res.* **2001**, *25*, 67–75. [CrossRef]
43. Misra, S.; Owen, R.; Kerr, J.; Greaves, M.; Elderfield, H. Determination of δ11B by HR-ICP-MS from mass limited samples: Application to natural carbonates and water samples. *Geochim. Cosmochim. Acta* **2014**, *140*, 531–552. [CrossRef]
44. Guillermic, M.; Misra, S.; Eagle, R.; Villa, A.; Chang, F.; Tripati, A. Seawater pH reconstruction using boron isotopes in multiple planktonic foraminifera species with different depth habitats and their potential to constrain pH and pCO_2 gradients. *Biogeosciences* **2020**, *17*, 3487–3510. [CrossRef]

45. Lloyd, N.S.; Sadekov, A.Y.; Misra, S. Application of 10^{13} ohm Faraday cup current amplifiers for boron isotopic analyses by solution mode and laser ablation multicollector inductively coupled plasma mass spectrometry. *Rapid Commun. Mass Spectrom.* **2017**, *32*, 9–18. [CrossRef]
46. McCulloch, M.T.; Holcomb, M.; Rankenburg, K.; Trotter, J.A. Rapid, high-precision measurements of boron isotopic compositions in marine carbonates. *Rapid Commun. Mass Spectrom.* **2014**, *28*, 2704–2712. [CrossRef]
47. Holcomb, M.; Venn, A.A.; Tambutté, E.; Allemand, D.; Trotter, J.; McCulloch, M.; Tambutte, S. Coral calcifying fluid pH dictates response to ocean acidification. *Sci. Rep.* **2014**, *4*, 5207. [CrossRef] [PubMed]
48. Sutton, J.N.; Liu, Y.-W.; Ries, J.B.; Guillermic, M.; Ponzevera, E.; Eagle, R.A. δ11B as monitor of calcification site pH in divergent marine calcifying organisms. *Biogeosciences* **2018**, *15*, 1447–1467. [CrossRef]
49. Gutjahr, M.; Bordier, L.; Douville, E.; Farmer, J.; Foster, G.L.; Hathorne, E.C.; Hönisch, B.; Lemarchand, D.; Louvat, P.; McCulloch, M.; et al. Sub-Permil Interlaboratory Consistency for Solution-Based Boron Isotope Analyses on Marine Carbonates. *Geostand. Geoanal. Res.* **2021**, *45*, 59–75. [CrossRef]
50. Hemming, N.; Hanson, G. Boron isotopic composition and concentration in modern marine carbonates. *Geochim. Cosmochim. Acta* **1992**, *56*, 537–543. [CrossRef]
51. Decarlo, T.M.; Comeau, S.; Cornwall, C.E.; McCulloch, M.T. Coral resistance to ocean acidification linked to increased calcium at the site of calcification. *Proc. R. Soc. B Boil. Sci.* **2018**, *285*, 20180564. [CrossRef]
52. Foster, G.L.; von Strandmann, P.P.; Rae, J. Boron and magnesium isotopic composition of seawater. *Geochem. Geophys. Geosyst.* **2010**, *11*. [CrossRef]
53. Dickson, A.G. Thermodynamics of the dissociation of boric acid in synthetic seawater from 273.15 to 318.15 K. *Deep Sea Res. Part A Oceanogr. Res. Pap.* **1990**, *37*, 755–766. [CrossRef]
54. Lee, K.; Kim, T.-W.; Byrne, R.H.; Millero, F.J.; Feely, R.A.; Liu, Y.-M. The universal ratio of boron to chlorinity for the North Pacific and North Atlantic oceans. *Geochim. Cosmochim. Acta* **2010**, *74*, 1801–1811. [CrossRef]
55. Upadhyay, D.; Lucarelli, J.; Arnold, A.; Flores, R.; Bricker, H.; Ulrich, R.N.; Jesmok, G.; Santi, L.; Defliese, W.; Eagle, R.A.; et al. Carbonate clumped isotope analysis (Δ_{47}) of 21 carbonate standards determined via gas-source isotope-ratio mass spectrometry on four instrumental configurations using carbonate-based standardization and multiyear data sets. *Rapid Commun. Mass Spectrom.* **2021**, *35*, e9143. [CrossRef] [PubMed]
56. Tukey, J. *Exploratory Data Analysis*; Addison-Wesley: Reading, MA, USA, 1977.
57. Hönisch, B.; Hemming, N.; Grottoli, A.; Amat, A.; Hanson, G.; Bijma, J. Assessing scleractinian corals as recorders for paleo-pH: Empirical calibration and vital effects. *Geochim. Cosmochim. Acta* **2004**, *68*, 3675–3685. [CrossRef]
58. Reynaud, S.; Hemming, N.G.; Juillet-Leclerc, A.; Gattuso, J.-P. Effect of pCO_2 and temperature on the boron isotopic composition of the zooxanthellate coral *Acropora* sp. *Coral Reefs* **2004**, *23*, 539–546. [CrossRef]
59. Trotter, J.; Montagna, P.; McCulloch, M.; Silenzi, S.; Reynaud, S.; Mortimer, G.; Martin, S.; Ferrier-Pagès, C.; Gattuso, J.-P.; Rodolfo-Metalpa, R. Quantifying the pH 'vital effect' in the temperate zooxanthellate coral Cladocora caespitosa: Validation of the boron seawater pH proxy. *Earth Planet. Sci. Lett.* **2011**, *303*, 163–173. [CrossRef]
60. Rollion-Bard, C.; Blamart, D.; Trebosc, J.; Tricot, G.; Mussi, A.; Cuif, J.-P. Boron isotopes as pH proxy: A new look at boron speciation in deep-sea corals using 11B MAS NMR and EELS. *Geochim. Cosmochim. Acta* **2011**, *75*, 1003–1012. [CrossRef]
61. Ries, J.B. A physicochemical framework for interpreting the biological calcification response to CO_2-induced ocean acidification. *Geochim. Cosmochim. Acta* **2011**, *75*, 4053–4064. [CrossRef]
62. Anagnostou, E.; Huang, K.-F.; You, C.-F.; Sikes, E.; Sherrell, R. Evaluation of boron isotope ratio as a pH proxy in the deep sea coral Desmophyllum dianthus: Evidence of physiological pH adjustment. *Earth Planet. Sci. Lett.* **2012**, *349–350*, 251–260. [CrossRef]
63. McCulloch, M.; Falter, J.; Trotter, J.; Montagna, P. Coral resilience to ocean acidification and global warming through pH up-regulation. *Nat. Clim. Chang.* **2012**, *2*, 623–627. [CrossRef]
64. Tanaka, K.; Holcomb, M.; Takahashi, A.; Kurihara, H.; Asami, R.; Shinjo, R.; Sowa, K.; Rankenburg, K.; Watanabe, T.; McCulloch, M. Response of Acropora digitifera to ocean acidification: Constraints from δ11B, Sr, Mg, and Ba compositions of aragonitic skeletons cultured under variable seawater pH. *Coral Reefs* **2015**, *34*, 1139–1149. [CrossRef]
65. Georgiou, L.; Falter, J.; Trotter, J.; Kline, D.I.; Holcomb, M.; Dove, S.G.; Hoegh-Guldberg, O.; McCulloch, M. pH homeostasis during coral calcification in a free ocean CO_2 enrichment (FOCE) experiment, Heron Island reef flat, Great Barrier Reef. *Proc. Natl. Acad. Sci. USA* **2015**, *112*, 13219–13224. [CrossRef]
66. Cai, W.J.; Ma, Y.; Hopkinson, B.M.; Grottoli, A.G.; Warner, M.E.; Ding, Q.; Hu, X.; Yuan, X.; Schoepf, V.; Xu, H.; et al. Microelectrode characterization of coral daytime interior pH and carbonate chemistry. *Nat. Commun.* **2016**, *7*, 11144. [CrossRef]
67. Comeau, S.; Cornwall, C.E.; McCulloch, M. Decoupling between the response of coral calcifying fluid pH and calcification to ocean acidification. *Sci. Rep.* **2017**, *7*, 7573. [CrossRef]
68. Comeau, S.; Tambutté, E.; Carpenter, R.C.; Edmunds, P.J.; Evensen, N.R.; Allemand, D.; Ferrier-Pagès, C.; Venn, A.A. Coral calcifying fluid pH is modulated by seawater carbonate chemistry not solely seawater pH. *Proc. R. Soc. B Boil. Sci.* **2017**, *284*, 20161669. [CrossRef] [PubMed]
69. Liu, Y.-W.; Sutton, J.N.; Ries, J.B.; Eagle, R.A. Regulation of calcification site pH is a polyphyletic but not always governing response to ocean acidification. *Sci. Adv.* **2020**, *6*, eaax1314. [CrossRef]

70. Allison, N.; Cole, C.; Hintz, C.; Hintz, K.; Rae, J.; Finch, A. Resolving the interactions of ocean acidification and temperature on coral calcification media pH. *Coral Reefs* **2021**, *40*, 1807–1818. [CrossRef]
71. Allison, N.; Cohen, I.; Finch, A.A.; Erez, J. Controls on Sr/Ca and Mg/Ca in scleractinian corals: The effects of Ca-ATPase and transcellular Ca channels on skeletal chemistry. *Geochim. Cosmochim. Acta* **2011**, *75*, 6350–6360. [CrossRef]
72. Pierrot, D.; Lewis, E.; Wallace, D.W.R. MS Excel Program Developed for CO_2 System Calculations. In *Carbon Dioxide Information Analysis Center, Oak Ridge National Laboratory*; U.S. Department of Energy: Oak Ridge, TN, USA, 2006. [CrossRef]
73. Al-Horani, F.A.; Almoghrabi, S.M.; De Beer, D. The mechanism of calcification and its relation to photosynthesis and respiration in the scleractinian coral Galaxea fascicularis. *Mar. Biol.* **2003**, *142*, 419–426. [CrossRef]
74. McCulloch, M.T.; D'Olivo, J.P.; Falter, J.; Holcomb, M.; Trotter, J. Coral calcification in a changing World and the interactive dynamics of pH and DIC upregulation. *Nat. Commun.* **2017**, *8*, 15686. [CrossRef] [PubMed]
75. Allison, N. Reconstructing coral calcification fluid dissolved inorganic carbon chemistry from skeletal boron: An exploration of potential controls on coral aragonite B/Ca. *Heliyon* **2017**, *3*, e00387. [CrossRef] [PubMed]
76. Mavromatis, V.; Montouillout, V.; Noireaux, J.; Gaillardet, J.; Schott, J. Characterization of boron incorporation and speciation in calcite and aragonite from co-precipitation experiments under controlled pH, temperature and precipitation rate. *Geochim. Cosmochim. Acta* **2015**, *150*, 299–313. [CrossRef]
77. Holcomb, M.; DeCarlo, T.; Gaetani, G.; McCulloch, M. Factors affecting B/Ca ratios in synthetic aragonite. *Chem. Geol.* **2016**, *437*, 67–76. [CrossRef]
78. Mass, T.; Giuffre, A.J.; Sun, C.-Y.; Stifler, C.A.; Frazier, M.J.; Neder, M.; Tamura, N.; Stan, C.V.; Marcus, M.A.; Gilbert, P.U.P.A. Amorphous calcium carbonate particles form coral skeletons. *Proc. Natl. Acad. Sci. USA* **2017**, *114*, E7670–E7678. [CrossRef]
79. Evans, D.; Gray, W.R.; Rae, J.; Greenop, R.; Webb, P.B.; Penkman, K.; Kröger, R.; Allison, N. Trace and major element incorporation into amorphous calcium carbonate (ACC) precipitated from seawater. *Geochim. Cosmochim. Acta* **2020**, *290*, 293–311. [CrossRef]
80. Ulrich, R.N.; Guillermic, M.; Campbell, J.; Hakim, A.; Han, R.; Singh, S.; Stewart, J.D.; Román-Palacios, C.; Carroll, H.M.; De Corte, I.; et al. Patterns of Element Incorporation in Calcium Carbonate Biominerals Recapitulate Phylogeny for a Diverse Range of Marine Calcifiers. *Front. Earth Sci.* **2021**, *9*, 641760. [CrossRef]
81. Giuffre, A.J.; Gagnon, A.C.; De Yoreo, J.J.; Dove, P.M. Isotopic tracer evidence for the amorphous calcium carbonate to calcite transformation by dissolution–reprecipitation. *Geochim. Cosmochim. Acta* **2015**, *165*, 407–417. [CrossRef]
82. Von Euw, S.; Zhang, Q.; Manichev, V.; Murali, N.; Gross, J.; Feldman, L.C.; Gustafsson, T.; Flach, C.; Mendelsohn, R.; Falkowski, P.G. Biological control of aragonite formation in stony corals. *Science* **2017**, *356*, 933–938. [CrossRef]
83. Swart, P.; Leder, J.; Szmant, A.; Dodge, R. The origin of variations in the isotopic record of scleractinian corals: II. Carbon. *Geochim. Cosmochim. Acta* **1996**, *60*, 2871–28856. [CrossRef]
84. Grottoli, A.G. Effect of light and brine shrimp on skeletal δ13C in the Hawaiian coral Porites compressa: A tank experiment. *Geochim. Cosmochim. Acta* **2002**, *66*, 1955–1967. [CrossRef]
85. Baker, A.C. Reef corals bleach to survive change. *Nature* **2001**, *411*, 765–766. [CrossRef] [PubMed]
86. Baker, D.M.; Freeman, C.J.; Wong, J.C.; Fogel, M.L.; Knowlton, N. Climate change promotes parasitism in a coral symbiosis. *ISME J.* **2018**, *12*, 921–930. [CrossRef] [PubMed]
87. Castillo, K.D.; Ries, J.B.; Bruno, J.F.; Westfield, I.T. The reef-building coral *Siderastrea siderea* exhibits parabolic responses to ocean acidification and warming. *Proc. R. Soc. B Boil. Sci.* **2014**, *281*, 20141856. [CrossRef] [PubMed]
88. Okazaki, R.R.; Towle, E.K.; Hooidonk, R.; Mor, C.; Winter, R.N.; Piggot, A.M.; Cunning, R.; Baker, A.; Klaus, J.S.; Swart, P.K.; et al. Species-specific responses to climate change and community composition determine future calcification rates of Florida Keys reefs. *Glob. Chang. Biol.* **2016**, *23*, 1023–1035. [CrossRef]
89. Bove, C.B.; Umbanhowar, J.; Castillo, K.D. Meta-Analysis Reveals Reduced Coral Calcification Under Projected Ocean Warming but Not Under Acidification Across the Caribbean Sea. *Front. Mar. Sci.* **2020**, *7*, 127. [CrossRef]

Journal of
Marine Science and Engineering

MDPI

Article

Impacts of Warming and Acidification on Coral Calcification Linked to Photosymbiont Loss and Deregulation of Calcifying Fluid pH

Louise P. Cameron [1,2], Claire E. Reymond [2,3], Jelle Bijma [4], Janina V. Büscher [5], Dirk De Beer [6], Maxence Guillermic [7], Robert A. Eagle [7], John Gunnell [1], Fiona Müller-Lundin [1], Gertraud M. Schmidt-Grieb [5], Isaac Westfield [1], Hildegard Westphal [2,8,9] and Justin B. Ries [1,2,*]

1 Department of Marine and Environmental Sciences, Northeastern University, Nahant, MA 01908, USA
2 Leibniz Centre for Tropical Marine Research (ZMT), 28359 Bremen, Germany
3 Geocoastal Research Group, School of Geosciences, The University of Sydney, Camperdown, Sydney, NSW 2006, Australia
4 Alfred Wegener Institute, Helmholtz Center for Polar and Marine Research, Am Handelshafen 12, 27570 Bremerhaven, Germany
5 GEOMAR Helmholtz Centre for Ocean Research Kiel, 24148 Kiel, Germany
6 Max Planck Institute for Marine Microbiology, 28359 Bremen, Germany
7 Atmospheric and Oceanic Sciences Department, Institute of the Environment and Sustainability, Earth, Planetary and Space Sciences, University of California, Los Angeles, CA 90095, USA
8 Department for Geosciences, University of Bremen, 28359 Bremen, Germany
9 Physical Science and Engineering Division, King Abdullah University of Science and Technology (KAUST), Thuwal, Makkah 23955-6900, Saudi Arabia
* Correspondence: j.ries@northeastern.edu

Citation: Cameron, L.P.; Reymond, C.E.; Bijma, J.; Büscher, J.V.; De Beer, D.; Guillermic, M.; Eagle, R.A.; Gunnell, J.; Müller-Lundin, F.; Schmidt-Grieb, G.M.; et al. Impacts of Warming and Acidification on Coral Calcification Linked to Photosymbiont Loss and Deregulation of Calcifying Fluid pH. *J. Mar. Sci. Eng.* **2022**, *10*, 1106. https://doi.org/10.3390/jmse10081106

Academic Editor: Weidong Zhai

Received: 20 July 2022
Accepted: 6 August 2022
Published: 12 August 2022

Abstract: Corals are globally important calcifiers that exhibit complex responses to anthropogenic warming and acidification. Although coral calcification is supported by high seawater pH, photosynthesis by the algal symbionts of zooxanthellate corals can be promoted by elevated pCO_2. To investigate the mechanisms underlying corals' complex responses to global change, three species of tropical zooxanthellate corals (*Stylophora pistillata*, *Pocillopora damicornis*, and *Seriatopora hystrix*) and one species of asymbiotic cold-water coral (*Desmophyllum pertusum*, syn. *Lophelia pertusa*) were cultured under a range of ocean acidification and warming scenarios. Under control temperatures, all tropical species exhibited increased calcification rates in response to increasing pCO_2. However, the tropical species' response to increasing pCO_2 flattened when they lost symbionts (i.e., bleached) under the high-temperature treatments—suggesting that the loss of symbionts neutralized the benefit of increased pCO_2 on calcification rate. Notably, the cold-water species that lacks symbionts exhibited a negative calcification response to increasing pCO_2, although this negative response was partially ameliorated under elevated temperature. All four species elevated their calcifying fluid pH relative to seawater pH under all pCO_2 treatments, and the magnitude of this offset ($\Delta[H^+]$) increased with increasing pCO_2. Furthermore, calcifying fluid pH decreased along with symbiont abundance under thermal stress for the one species in which calcifying fluid pH was measured under both temperature treatments. This observation suggests a mechanistic link between photosymbiont loss ('bleaching') and impairment of zooxanthellate corals' ability to elevate calcifying fluid pH in support of calcification under heat stress. This study supports the assertion that thermally induced loss of photosymbionts impairs tropical zooxanthellate corals' ability to cope with CO_2-induced ocean acidification.

Keywords: microelectrode; ocean acidification; global warming; calcifying fluid; scleractinian coral; zooxanthellate photosymbiont; photosynthesis; calcification; bleaching

1. Introduction

Anthropogenic emissions are predicted to cause sea-surface warming [1] and ocean acidification (OA)—a process that lowers seawater pH and aragonite saturation state [2] (Ω_A). OA increases both the dissolution rate of $CaCO_3$ shell/skeleton [3] and the rate at which new shell/skeleton is formed [4]. Tropical scleractinian corals are carbonate producers [5] that acquire nourishment via symbiotic photosynthetic zooxanthellae and from heterotrophic feeding [6]. They are vulnerable to warming as many exist near the upper end of their thermal tolerance limits [7]. Warming beyond a coral's thermal tolerance may cause the loss of photosymbionts—a process known as bleaching.

Coral calcification occurs in the coral's calcifying fluid, which is influenced by both external seawater chemistry and the coral itself [8]. Corals elevate saturation state of this fluid through a combination of pH elevation via the active removal of protons using membrane-bound Ca^{2+}-ATPase proton pumps [8–14] and DIC elevation [15,16]. Photosynthesis may aid pH elevation by supplying the necessary ATP required to drive Ca^{2+}-ATPase proton pumps. Under ocean acidification, the formation of $CaCO_3$ is more energetically costly [11,17,18], although this energetic cost is negligible compared to the total proportion of energy produced through photosynthesis [8]. Photosymbionts may therefore play a crucial role in mitigating the impacts of OA on corals by providing enough ATP to increase ion pumping rates in support of calcification. Photosynthesis may also be responsible for metabolic DIC elevation at the site of calcification, as implied by long-term seasonal variations in the concentration of DIC in the calcifying fluid [16]. Under warming, bleaching may impair coral calcification by reducing the amount of photosynthate that is translocated to the coral host, thereby increasing the coral's reliance on heterotrophic feeding [6], decreasing the metabolically derived DIC pool, and, potentially, reducing the amount of ATP available for proton-regulation in support of calcification.

Although tropical scleractinian corals generally exhibit parabolic calcification responses to ocean warming that are centered on their thermal optima [19], their calcification responses to OA are more nuanced. Many species exhibit reduced calcification rates [20,21], while others exhibit threshold responses [22], parabolic responses [19,21], or no calcification response to acidification [13,23,24].

This highlights the complexity of coral biomineralization, which can be biologically mediated by the secretion of skeletal organic molecules (SOM) such as adhesion, signaling, and structural proteins (e.g., calmodulin and sulphated acidic proteoglycans) [25–27]. The abundance of these SOMs in the calcifying fluid has been shown to alter the rate and morphology of aragonite precipitated [28–32].

Cold-water corals that inhabit deeper aphotic environments generally lack zooxanthellae and acquire nourishment exclusively through heterotrophic feeding [33]. The systems for regulating calcifying fluid pH (pH_{CF}) within azooxanthellate corals may therefore differ from zooxanthellate corals. Azooxanthellate corals exhibit reduced skeletal density [34], reduced calcification rate [35], and altered rates of respiration and feeding [36] in response to OA, but increased calcification rates under elevated temperature [35,37].

Numerous studies have investigated the isolated effects of ocean acidification (e.g., [11,24,38,39]) and thermal stress (e.g., [40]) on coral pH_{CF}. However, the present study—along with its companion paper [41]—are the first to investigate both the independent and combined effects of warming and acidification on coral pH_{CF}. These empirical constraints, combined with photosymbiont and calcification data for the three tropical coral species, yield insight into the mechanism by which warming and acidification interact to so negatively impact coral growth. Examining the roles that pH_{CF} regulation and symbiont abundance play in the coral calcification response to OA and warming should improve understanding and prediction of how different species of corals will respond to future global change.

We investigate these relationships by culturing three species of tropical zooxanthellate corals (*Stylophora pistillata*, *Pocillopora damicornis*, *Seriatopora hystrix*) and one azooxanthellate cold-water species (*Desmophyllum pertusum*, syn. *Lophelia pertusa*) under control (tropical = 28 °C, cold-water = 9 °C) and elevated (tropical = 31 °C, cold-water = 12 °C) temperatures at near-present-day (451–499 ppm), year 2100 (885–1096 ppm), and year 2500 (2807–3194 ppm) pCO_2 scenarios [1]. Coral calcifying fluid pH was measured with proton-sensitive microelectrodes. Symbiont abundance of the zooxanthellate corals was estimated based on coral color to investigate the role of photosymbionts in the coral calcification response to warming and acidification.

This work is a companion paper to Guillermic et al. [41], which investigated the impacts of pCO_2 and temperature on the calcifying fluid chemistry and calcification rate of two tropical species of scleractinian corals, *S. pistillata* and *P. damicornis*. The present study builds upon its companion paper by investigating these relationships for two additional species of scleractinian corals—the tropical species *S. hystrix* and the cold-water species *L. pertusa*. The present study also includes photosymbiont data for all three tropical coral species cultured under all experimental treatments.

2. Materials and Methods

2.1. Overview of Experimental Design

Three species of tropical scleractinian corals (*S. pistillata*, *P. damicornis*, and *S. hystrix*) were cultured under three ocean acidification (OA) scenarios established by modification of pCO_2 at control and elevated temperatures (control acidification: 28.29 °C ± 0.01 s.e./466 ppm ± 8; 31.72 °C ± 0.06/499 ppm ± 9; moderate acidification: 27.88 °C ± 0.02/925 ppm ± 15; 30.83 °C ± 0.02/885 ppm ± 12; high acidification: 28.17 °C ± 0.02/2807 ppm ± 119; 30.93 °C ± 0.02/3194 ppm ± 135) in four replicate tanks at each treatment for 62 days in April–June 2016 (Table 1). The control temperature treatment was assigned to fall within the natural temperature range of coral reefs in Fiji (26–29 °C), where these corals were collected [42]. Likewise, the elevated temperature was assigned to be just above the bleaching threshold (30–30.5 °C) for corals in Fiji [42]. Simultaneously, the cold-water coral *L. pertusa* was cultured under three OA scenarios at control and elevated temperatures (control acidification: 8.86 °C ± 0.02/451 ppm ± 24; 12.17 °C ± 0.03/494 ppm ± 16; moderate acidification: 9.07 °C ± 0.03/1096 ppm ± 99; 12.52 °C ± 0.04/1079 ppm ± 60; high acidification: 8.83 °C ± 0.02/2864 ppm ± 222; 12.67 °C ± 0.03/3167 ppm ± 202) in four replicate tanks at each treatment for 33 days in April–May 2016 (Table 2). Corals were acclimated to laboratory conditions for one week and then to experimental conditions for an additional two weeks prior to the start of the experiment. Calcification rates, pH_{CF}, and relative photosymbiont abundance were quantified during the experiment (details below).

2.2. Coral Husbandry

Experiments were carried out in the MAREE marine experimental facility at the Leibniz Centre for Tropical Marine Research (ZMT). Specimens of the tropical scleractinian coral species *S. pistillata* and *P. damicornis* were obtained from DeJong MarineLife (Netherlands). Colony-level information was not available for these specimens. Specimens of the tropical scleractinian coral species *S. hystrix* were obtained from an experimental stock colony provided by the ZMT. Fragments of the cold-water coral *L. pertusa* were obtained from four colonies of an experimental stock culture provided by the Marine Biogeochemistry Department of the Helmholtz Centre for Ocean Research in Kiel (GEOMAR), previously collected at a depth of ca. 200 m at Nord-Leksa Reef in Trondheimsfjord, Norway. Coral fragments (*S. pistillata* = 65, *P. damicornis* = 63, *S. hystrix* = 52, and *L. pertusa* = 43) were mounted onto 3 cm x 3 cm plastic egg-crate stands using cyanoacrylate epoxy and assigned a unique identifier. Each treatment was replicated in four tanks. Equivalent size ranges of specimens were maintained across treatments. Coral specimens obtained from larger colonies were randomly distributed amongst pCO_2 and temperature treatments and replicate tanks. Coral specimens that died before the completion of the experiment were promptly removed from the tanks so as not to impact the remaining live corals in the experiment. Additional details about the number and weight of individuals in each treatment are provided in Section S1 of the Supplementary Online Material.

Table 1. Average calculated parameters for the tropical corals and all treatments: pCO_2 of the mixed gases in equilibrium with seawaters (pCO_2 (gas-e)), pH on seawater scale (pH_{SW}), carbonate ion concentration ($[CO_3^{2-}]$), bicarbonate ion concentration ($[HCO_3^-]$), dissolved carbon dioxide ($[CO_2]_{SW}$), and aragonite saturation state (Ω_A). Average measured parameters for all treatments: salinity (Sal), temperature (Temp), pH on NBS scale (pH_{NBS}), total alkalinity (TA), and dissolved inorganic carbon (DIC). 'SE' represents standard error and 'n' is the sample size.

		400 ppm (9 °C)	400 ppm (12 °C)	1000 ppm (9 °C)	1000 ppm (12 °C)	2800 ppm (9 °C)	2800 ppm (12 °C)
		CALCULATED PARAMETERS					
$pCO_{2\,(gas\text{-}e)}$	(ppm-v)	466	499	925	885	2807	3194
	SE	8	9	15	12	119	135
	Range	362–540	425–607	808–1144	772–1050	1728–4302	2298–4945
	n	32	32	31	32	32	32
pH_{SW}		8.11	8.06	7.85	7.87	7.42	7.42
	SE	0.02	0.02	0.01	0.01	0.01	0.02
	Range	8.03–8.30	7.97–8.27	7.73–7.98	7.80–7.99	7.28–7.51	7.20–7.57
	n	32	31	31	32	32	32
$[CO_3^{2-}]$	(μM)	334	320	217	265	90	113
	SE	9	11	6	4	3	4
	Range	235–395	211–442	159–274	226–309	64–113	76–149
	n	32	32	31	32	32	32
$[HCO_3^-]$	(μM)	2133	2017	2424	2468	2697	3030
	SE	33	36	35	35	77	54
	Range	1846–2433	1772–2389	2138–2748	2193–2842	2104–3289	2476–3475
	n	32	31	31	32	32	32
$[CO_2]_{(SW)}$	(μM)	12.3	12.3	24.6	22.0	74.3	79.5
	SE	0.2	0.3	0.4	0.4	3.2	3.4
	Range	9–14	10–17	21–30	19–27	45–112	56–121
	n	32	32	31	32	32	32
Ω_A		5.4	5.2	3.5	4.3	1.4	1.8
	SE	0.1	0.2	0.1	0.1	0.0	0.1
	Range	3.8–6.4	3.5–7.2	2.6–4.4	3.7–5.0	1.0–1.8	1.2–2.4
	n	32	32	31	32	32	32
		MEASURED PARAMETERS					
Sal	(psu)	34.87	35.63	35.44	35.99	35.75	35.68
	SE	0.06	0.08	0.05	0.06	0.05	0.06
	Range	33.75–36.25	34.45–37.55	34.55–36.65	34.85–37.25	34.55–36.75	34.35–37.05
	n	104	104	104	104	104	104
Temp	(°C)	28.29	31.72	27.88	30.83	28.17	30.93
	SE	0.01	0.06	0.02	0.02	0.02	0.02
	Range	28.1–28.5	30.8–32.3	27.6–28.2	30.5–31.0	28.0–28.4	30.6–31.2
	n	104	104	104	104	104	104
pH_{NBS}		8.27	8.24	8.04	8.12	7.62	7.69
	SE	0.01	0.01	0.01	0.01	0.01	0.01
	Range	8.02–8.47	8.12–8.48	7.88–8.33	7.92–8.34	7.47–7.98	7.48–7.97
	n	104	104	104	104	104	104
TA	(μM)	2915	2774	2939	3083	2907	3290
	SE	49	54	46	40	81	56
	Range	2420–3304	2309–3223	2524–3306	2735–3462	2260–3493	2705–3794
	n	32	32	32	32	32	32
DIC	(μM)	2480	2350	2645	2755	2861	3223
	SE	40	44	44	37	82	57
	Range	2097–2824	2009–2726	2013–3003	2445–3134	2231–3490	2634–3700
	n	32	32	32	32	32	32

Table 2. Average calculated parameters for *Lophelia pertusa* and all treatments: pCO_2 of the mixed gases in equilibrium with seawaters (pCO_2 (gas-e)), pH on seawater scale (pH_{SW}), carbonate ion concentration ($[CO_3^{2-}]$), bicarbonate ion concentration ($[HCO_3^-]$), dissolved carbon dioxide ($[CO_2]_{SW}$), and aragonite saturation state (Ω_A). Average measured parameters for all treatments: salinity (Sal), temperature (Temp), pH on NBS scale (pH_{NBS}), total alkalinity (TA), and dissolved inorganic carbon (DIC). 'SE' represents standard error and 'n' is the sample size.

		400 ppm (9 °C)	400 ppm (12 °C)	1000 ppm (9 °C)	1000 ppm (12 °C)	2800 ppm (9 °C)	2800 ppm (12 °C)
				CALCULATED PARAMETERS			
$pCO_{2\ (gas\text{-}e)}$	(ppm-v)	451	494	1096	1079	2864	3167
	SE	24	16	99	60	222	202
	Range	238–580	399–595	672–1815	834–1496	1600–4162	1821–4669
	n	16	16	16	16	16	16
pH_{SW}		8.20	8.21	7.87	7.86	7.42	7.42
	SE	0.04	0.04	0.04	0.03	0.04	0.03
	Range	8.01–8.54	7.96–8.40	7.62–8.12	7.70–8.06	7.17–7.67	7.19–7.63
	n	20	20	20	20	20	20
$[CO_3^{2-}]$	(μM)	244	289	121	140	52	60
	SE	23	25	9	8	6	5
	Range	128–486	149–458	87–201	102–211	25–99	35–110
	n	20	20	20	20	20	20
$[HCO_3^-]$	(μM)	2674	2775	2794	2835	3034	3058
	SE	107	111	89	95	69	81
	Range	2028–3356	2215–3348	2299–3366	2350–3349	2655–3420	2604–3486
	n	20	20	20	20	20	20
$[CO_2]_{(SW)}$	(μM)	20	19	45	41	127	118
	SE	1	1	4	2	8	8
	Range	11–26	14–24	28–83	30–59	72–190	72–187
	n	20	20	20	20	20	20
Ω_A		3.7	4.4	1.8	2.1	0.8	0.9
	SE	0.3	0.4	0.1	0.1	0.1	0.1
	Range	1.9–7.3	2.3–6.9	1.3–3.0	1.6–3.2	0.4–1.5	0.5–1.7
	n	20	20	20	20	20	20
				MEASURED PARAMETERS			
Sal	(psu)	34.98	35.13	34.96	35.23	35.01	35.40
	SE	0.02	0.03	0.03	0.06	0.02	0.05
	Range	34.55–35.25	34.65–35.65	34.55–35.65	34.20–36.15	34.45–35.25	34.35–36.05
	n	60	60	60	60	60	60
Temp	(°C)	8.86	12.17	9.07	12.52	8.83	12.67
	SE	0.03	0.04	0.03	0.04	0.02	0.03
	Range	8.5–9.4	11.6–12.6	8.7–9.4	12.0–12.9	8.5–9.1	12.0–13.1
	n	60	60	60	60	60	60
pH_{NBS}		8.07	8.13	7.80	7.82	7.39	7.40
	SE	0.01	0.02	0.02	0.02	0.02	0.02
	Range	7.82–8.39	7.85–8.43	7.48–8.15	7.52–8.01	7.13–7.75	7.12–7.63
	n	60	60	60	60	60	60
TA	(μM)	3245	3442	3080	3163	3157	3198
	SE	148	161	93	106	80	87
	Range	2355–4106	2601–4349	2598–3570	2609–3783	2752–3570	2759–3655
	n	20	20	20	20	20	20
DIC	(μM)	2938	3083	2960	3016	3214	3235
	SE	125	134	93	101	69	85
	Range	2180–3659	2395–3793	2449–3536	2494–3556	2805–3599	2741–3676
	n	20	20	20	20	20	20

After 1-week of acclimation at control conditions, temperature and pCO_2 were then incrementally increased to treatment levels over an additional week, after which time coral specimens were acclimated to treatment conditions for an additional week prior to the experiment. All coral specimens were cultured in 10-L replicate tanks supplied with seawater from 244-L sumps, where water filtration, temperature control, and pCO_2 control occurred. Seawater was filtered with protein skimmers, mechanical filters, and activated charcoal. All tropical coral aquaria were illuminated with 150 lux using actinic blue and white aquarium lights on a 12-h light/dark cycle. Aquaria holding the cold-water coral *L. pertusa* were not illuminated as this species lives below the photic zone. Each experimental treatment containing *L. pertusa* specimens (comprised of 4 replicate tanks) shared a water source with a separate reservoir containing five specimens of the king scallop *Pecten maximus* cultured as part of a separate experiment [43]. Seawater was filtered with protein skimmers, mechanical filters, and activated charcoal before returning to the *L. pertusa* tanks.

During acclimation and experimental periods, the tropical corals were fed 1-day old *Artemia salina* nauplii hatched from ca. 40 mg of eggs. Approximately 10 mL of concentrated live nauplii were introduced into each replicate tank every second day. The food mixture was pipetted directly adjacent to each coral specimen. Specimens of *L. pertusa* were fed the same diet supplemented with 20 mL of *Calanus finmarchicus* concentrate (Goldpods) suspended in the initial aliquot of *Artemia salina*. Corals were fed at the end of the day, and all filtration material was cleaned the following morning.

2.3. Seawater Chemistry Manipulation and Measurement

Measured and calculated carbonate system parameters are summarized in Tables 1 and 2. Experimental tank pCO_2 was maintained by vigorously bubbling mixtures of CO_2-free air and CO_2 into the 244-L treatment sumps with microporous sparging tubes. The pCO_2 of the bubbled gases was achieved by mixing compressed CO_2-free air and compressed CO_2 with solenoid-valve mass flow controllers at flow rates proportional to the target pCO_2 conditions. Natural seawater, originally collected from Spitsbergen, Norway, was continuously added to each of the 244-L sumps at a rate of 0.6 L/hour. Temperature (s.e.) for the tropical corals was maintained at 28 (0.02) °C and 31 (0.06) °C using 125-watt aquarium heaters (EHEIM), controlled with a programmable thermostat. Temperature for *L. pertusa* was maintained at 9 °C (0.03) and 12 °C (0.04) using aquarium chillers (Aqua Medic).

Temperature, pH, and salinity of all replicate tanks were measured three times per week using a multi-electrode probe (WTW Multi 3430 Set K). Samples for the analysis of dissolved inorganic carbon (DIC) and total alkalinity (TA) were collected weekly from each of the replicate tanks at midday and used to calculate other carbonate system parameters using the program CO2SYS [44]. Nutrient concentrations ($[NO_3^-]$, $[PO_4^{3-}]$, and $[NH_4^+]$) of all replicate tanks were measured weekly. Additional details about the methods used to measure the carbonate system and nutrient concentrations of replicate tanks are provided in Section S2 of the Supplementary Online Materials.

2.4. Calcification Rates

Calcification rates were calculated from the change in estimated dry weight of all coral fragments over the experimental period. Dry weights were estimated from buoyant weight measurements taken at the start and end of the experimental period according to the following empirically derived relationships:

Stylophora pistillata: Dry weight (g) = 1.919 × Buoyant Weight (g) + 7.677;

Pocillopora damicornis: Dry weight (g) = 1.662 × Buoyant Weight (g) + 8.777;

Seriatopora hystrix: Dry weight (g) = 1.668 × Buoyant Weight (g) + 8.493;

Lophelia pertusa: Dry weight (g) = 1.594 × Buoyant Weight (g) − 0.206;

where the precision of this relationship is equivalent to the standard error of the regression (*S. pistillata*: 0.060 g; *P. damicornis*: 0.053 g; *S. hystrix*: 0.068 g; *L. pertusa*: 0.023 g). Additional details about the methods used to calculate calcification rate via buoyant weights are provided in Section S3 of the Supplementary Online Materials.

The number of days between the start and end buoyant weight measurements was then used to standardize %-calcification to a daily rate. Coral skeletons were also labeled with the fluorescent dye calcein (30 mg Se-Mark liquid calcein/kg-seawater) for 5 days prior to the initial buoyant weighing to identify skeletal material produced exclusively under the experimental conditions. Although all four species of corals recorded the calcein marker in their coral skeleton, rates of linear extension could not be reliably measured from the calcein marker because the dye was not incorporated into the skeletons in a consistent manner (see Section S4 of the Supplementary Online Materials for images of coral uptake of the calcein dye).

2.5. Estimating Coral Photosymbiont Index

The tropical coral specimens were photographed alongside the Coral Watch Coral Health Chart color scale [45–47] (Section S5 of the Supplementary Online Materials) under 150 lux (i.e., equivalent lighting to their experimental treatments) at the end of the experimental period to estimate relative photosymbiont abundance (a proxy for bleaching) of the coral specimens. This method involved extracting red-band color of the live coral tissue and the color scale and then assigning each pixel within the coral tissue image a discrete score (1–5) relative to the red-band values of the color scale (Figure 1). Additional details about the methods used to process photographs used for the estimation of photosymbiont index are provided in Section S5 of the Supplementary Online Materials.

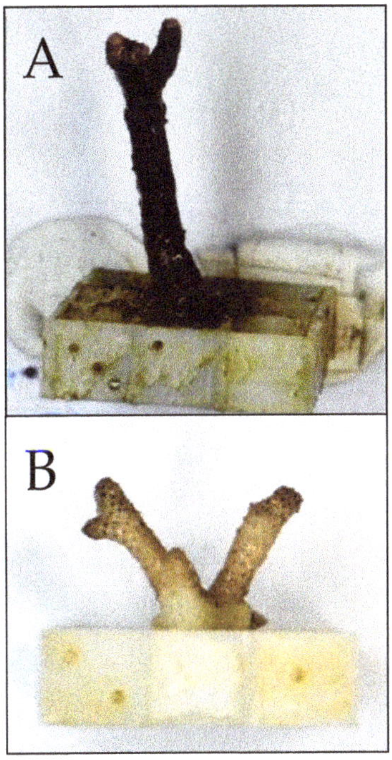

Figure 1. Representative images used in the estimation of relative photosymbiont abundance. Panel (**A**) depicts a healthy, unbleached coral from the control temperature, high pCO_2 treatment (color score = 5.69). Panel (**B**) depicts a partially bleached coral from the high temperature, high pCO_2 treatment (color score = 4.56).

2.6. Measurement of Calcifying Fluid pH

Calcifying fluid pH was measured using proton-sensitive liquid ion-exchanger (LIX) microelectrodes produced at the Max Planck Institute for Marine Microbiology (MPIMM) using a modified version of the technique described in De Beer et al. [48]. In brief, green soda lime glass microcapillary tubes (Schott model 8516) were held in a heated coil and pulled to a target tip diameter of ca. 10 μm, yielding final diameters of 8–20 μm. These were then silanized to produce a hydrophobic surface that allowed the adhesion of the LIX membrane. The microcapillary tubes were filled with ca. 300 μm of degassed, filtered electrolyte (300 mM KCl, 50 mM sodium phosphate adjusted to pH 7.0) using a plastic syringe with a 0.1-mm tip. The microcapillary tubes were then backfilled with LIX containing a polyvinyl chloride (PVC) epoxy to prevent leakage of electrolyte by submerging the tips of the microcapillary tubes in LIX and apply suction to the other end of the tube until the PVC-containing LIX was drawn into the tip of the microcapillary by 100–200 μm. Microcapillary tubes were encased in a Pasteur pipette for shielding, with the pulled tip of the microcapillary tube protruding ca. 2 cm beyond this casing. This casing was filled with a 0.3 M KCl solution and connected to the reference electrode with an Ag/AgCl wire to minimize electrical noise. Microelectrodes were left for 24 h after construction to allow for stabilization of the LIX membranes.

All microelectrode equipment (millivolt meter, National Instruments DAQ Pad 6020E, laptop, cables, micromanipulator, VT80 Micos motor arm, lab stands, Zeiss Stemi SV6 binocular microscope) was set up adjacent to the experimental tanks to minimize transport stress for the corals. Two reservoirs of seawater, sourced from the corresponding experimental treatment tanks, were established next to the microelectrode system. These reservoirs were bubbled with the corresponding treatment gases and maintained at the corresponding treatment temperature using aquarium heaters or chillers. The seawater was circulated between the two reservoirs through two 5.4 L flow-through chambers (30 × 12 × 15 cm). All pH microelectrode measurements were performed within these smaller flow-through chambers.

Measurements of calcifying fluid pH were made in the flow-through chambers filled with treatment seawater. Light levels in these chambers were measured using a digital lux-meter positioned next to the target coral polyp and were maintained at 150 lux. All corals were acclimated to the microelectrode chamber until polyp extension was observed prior to measurements (minimum of 10 min). Measurements of calcifying fluid pH were performed on three replicate individuals per treatment, with one measurement obtained for each individual. Calcifying fluid pH measurements were obtained for all species in all pCO_2 treatments under the control temperature treatment. Due to constraints on time and resources available for the experiment, calcifying fluid pH measurements under the high temperature treatment were only obtained for one species (*S. pistillata*).

The proton-sensitive LIX microelectrodes were used to measure both seawater and calcifying fluid pH. Before and after measurement of calcifying fluid pH, all microelectrodes were calibrated at the treatment temperature with pH 7 and 9 NBS buffers. The vertical position of the microelectrode was controlled with one-micron precision using a motorized micromanipulator. The microelectrodes were slowly inserted with a micromanipulator through the coral tissue into the upper portion of the coral calyx, between septal ridges and proximal to the thecal wall, until the skeleton was reached. This positioning of the electrode was verified by a shift in the pH profile [14] (Figure 2). A vertical pH profile (Figure 2) was then obtained by moving the microelectrode out of the calyx into the adjacent seawater. This profile was obtained in 1 μm steps for the first 20 μm, followed by 5 μm steps out into the surrounding seawater.

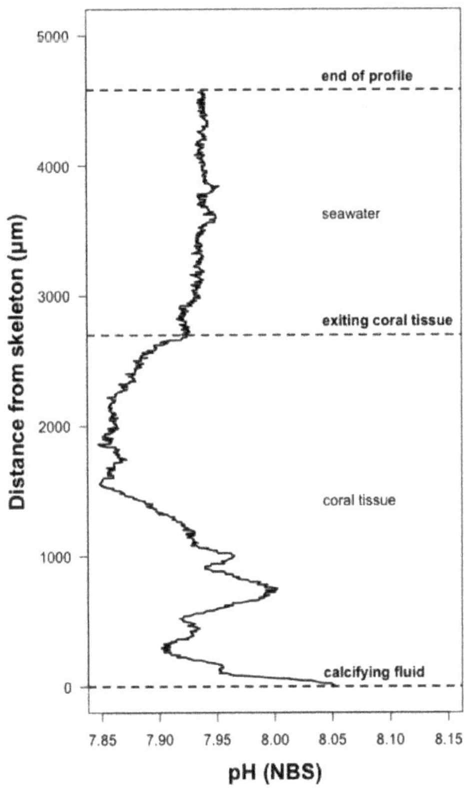

Figure 2. An example of a vertical pH profile generated during measurement of the coral calcifying fluid pH. This particular pH profile was generated for a specimen of *S. hystrix* cultured in the '1000 ppm pCO_2, 28 °C' treatment.

The 1-µm spatial resolution of the micromanipulator allowed for the positioning of the electrode within the thin calcifying fluid immediately adjacent to the coral skeleton. If the skeleton was inadvertently contacted during this positioning, it is possible that the tip of the microelectrode would break and render it dysfunctional. It was visually evident if the microelectrode tip broke upon contact with the skeleton, and this would also result in an abrupt voltage anomaly, often followed by a drift in the voltage even while the electrode was in a fixed position. The pH profile was aborted if there was evidence of any of these issues and then reinitiated with a new microelectrode.

The calibration and microelectrode pH data were processed by parsing scatter-plots of the data into three zones, which were annotated at the time of data collection. The calibration data were parsed as pH 7 buffer and pH 9 buffer. The microelectrode pH data were parsed as calcifying fluid, tissue, and seawater. Notes recorded during the original measurements were used to assist in identifying boundaries of adjacent zones. Measured mV within each zone of the calcifying fluid measurements were converted to pH using the calibration regression produced for each microelectrode. The $\Delta[H^+]$ was calculated for each measured coral as the difference between the proton concentration ($[H^+]$) of the coral's surrounding seawater and the $[H^+]$ of the coral's calcifying fluid, both measured with the calibrated, proton-sensitive LIX microelectrodes.

2.7. Statistical Analysis

Statistical analyses were carried out in *R*. Corals that did not survive the experimental period (see Section S1 of the Supplementary Online Material) were excluded from analyses. A series of linear mixed effects models (lmers) were performed to investigate the influence of seawater pCO_2 and temperature on coral physiology (calcification rate, calcifying fluid pH, $\Delta[H^+]$, photosymbiont index), with treatment tank specified as a blocking factor [49]. Akaike information criterion (AIC) was used to estimate the relative amount of information lost by any given model [50]. The final model was chosen based on the lowest AIC score (whereby a lower score reflects a better fitting model) and highest R^2, which reflects the goodness of fit (from 0 to 1, 1 being a perfect fit) (see Section S6 of the Supplementary Online Material for AIC model selection tables). The normality and homoscedasticity of the chosen tests were then analyzed using diagnostic plots (QQ-plot, residuals vs. fitted plot), and normality was tested using a Shapiro–Wilk test. Color scores were square-root transformed to meet the assumption of normality. If an interaction term was significant, the individual levels of that interaction were examined in their own linear mixed effects models to interpret main effects.

Analysis of co-variance (ANCOVA) was used to investigate the impacts of seawater pCO_2, temperature, and species on calcification rate, thereby allowing interspecific comparisons of the impacts of OA and warming on coral calcification rate. An ANCOVA was also used to make interspecific comparisons of the impact of OA on calcifying fluid pH and $\Delta[H^+]$ of different coral species. The latter analyses excluded individuals from the elevated temperature treatment, as calcifying fluid pH was only obtained for one of the three coral species in this treatment.

Linear mixed effects models were used to examine the impact of photosymbiont index, as a proxy for photosymbiont abundance, on calcification rate. Multiple linear models were generated, starting with the model that contained the most terms (i.e., modeling calcification rate as a function of seawater pCO_2, temperature, and photosymbiont index). Final interpretations of the data were based upon the models that maximized R^2 and minimized AIC (see Table S8). An alpha of 0.05 was used for all models, whereby any relationship with a p-value of <0.05 was deemed statistically significant.

3. Results

3.1. Predictors of Calcification Rate

The calcification rate of all coral species (Figure 3) was significantly impacted by the interaction between seawater pCO_2 and temperature (statistical significance indicated by p-value ≤ 0.05; Table S4). Calcification rate significantly increased with pCO_2 under control temperature for all three tropical species, but showed no change across pCO_2 treatments under elevated temperature for *S. pistillata* or *P. damicornis*, and decreased with increasing pCO_2 under elevated temperature for *S. hystrix*. Temperature had a significant negative effect on calcification rate in all tropical species (Table S4. Calcification rate of the cold-water azooxanthellate coral *L. pertusa* declined significantly with increasing pCO_2 at both temperatures, but showed a significantly stronger response to pCO_2 under the control temperature (Figure 3; Table S4).

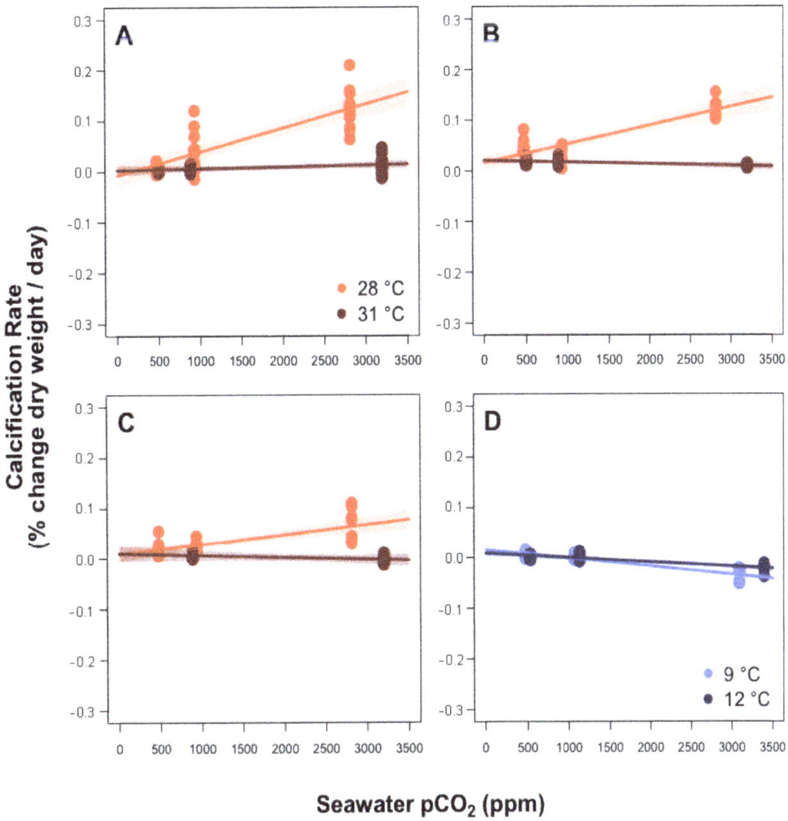

Figure 3. The relationship between pCO_2 and coral calcification rates at ambient and high temperature ((**A**) = *S. pistillata*, (**B**) = *S. hystrix*, (**C**) = *P. damicornis*, and (**D**) = *L. pertusa*). Shaded boundaries represent 95% confidence intervals. Calcification rate was significantly impacted by an interaction between pCO_2 and temperature for all coral species.

3.2. Calcifying Fluid Chemistry

The pH_{CF} of all four coral species was significantly greater than the pH of the corals' surrounding seawater (pH_{SW}) after 30 days of exposure to ocean acidification and warming (Figure 4A–D; lmer, *S. pistillata*: $p < 0.001$; *P. damicornis*: $p = 0.002$; *S. hystrix*: $p = 0.013$; *L. pertusa*: $p = 0.002$). The pH_{CF} of all four species declined significantly with increasing pCO_2 under control temperatures (Table S5), and, also, for *S. pistillata* under elevated temperature (*S. pistillata* was the only species for which pH_{CF} was measured at both control and elevated temperature; Figure 4). Under the elevated temperature treatment, pH_{CF} of *S. pistillata* remained higher than pH_{SW}, but was significantly lower than pH_{CF} at the control temperature (Table S5). The cold-water coral *Lophelia pertusa* exhibited the steepest decline in pH_{CF} with increasing pCO_2 (Figure 5).

All four coral species increased their $\Delta[H^+]$ (i.e., seawater $[H^+]$—calcifying fluid $[H^+]$) in response to increasing seawater pCO_2 (Figure 4E–H, Table S6). The $\Delta[H^+]$ of *S. pistillata* was significantly influenced by the interaction between pCO_2 and temperature (Table S6). In this species, there was no difference in $\Delta[H^+]$ between the two temperature treatments under control pCO_2. The $\Delta[H^+]$ increased with pCO_2 under both temperature treatments, but this increase was significantly greater in the control temperature treatment.

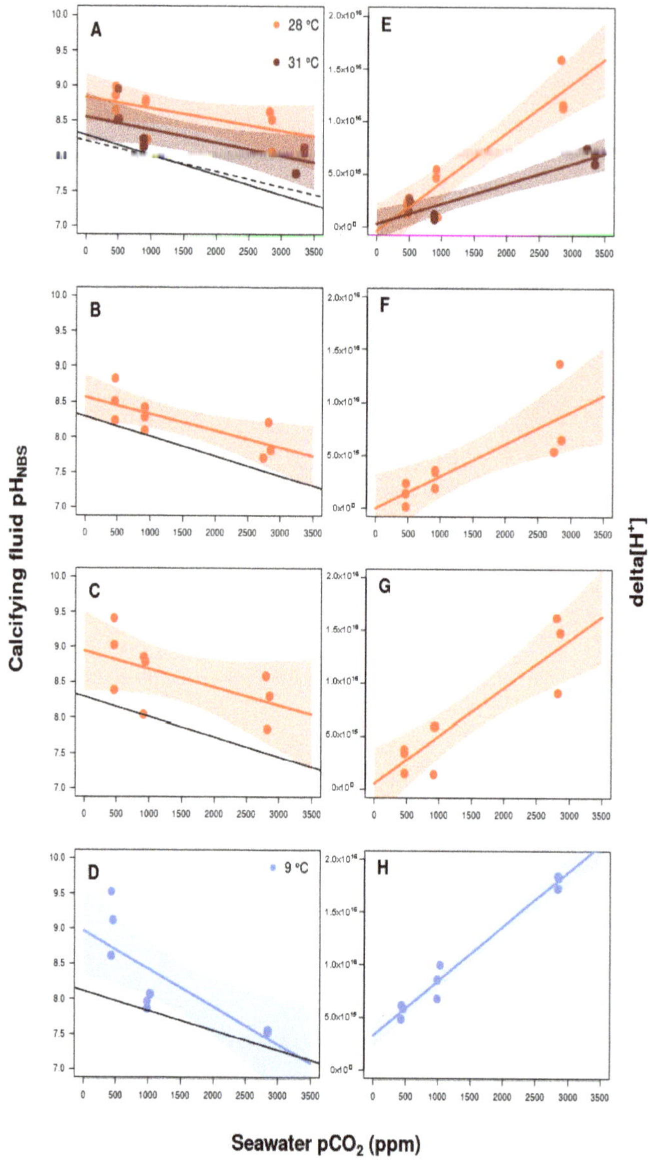

Figure 4. The effect of seawater pCO_2 on calcifying fluid pH (n = 3 individuals per treatment; panels (**A**–**D**)) and $\Delta[H^+]$ (panels (**E**–**H**)) for three species of tropical corals ((**A**,**E**) = *S. pistillata*; (**B**,**F**) = *S. hystrix*, and (**C**,**G**) = *P. damicornis*) and one species of cold-water coral (*L. pertusa*; (**D**,**H**)). The impact of elevated temperature on the response of *S. pistillata* to increasing pCO_2 is also shown (panels (**A**) and (**E**)). Increasing pCO_2 was significantly associated with declining calcifying fluid pH and increasing $\Delta[H^+]$ for all four coral species. The calcifying fluid pH of *S. pistillata* decreased significantly in response to a 3 °C increase in temperature, and $\Delta[H^+]$ significantly responded to an interaction between increased pCO_2 and increased temperature. Shaded boundaries represent 95% confidence intervals. Solid black lines represent seawater pH under control temperature; dashed black line represents seawater pH under high temperature.

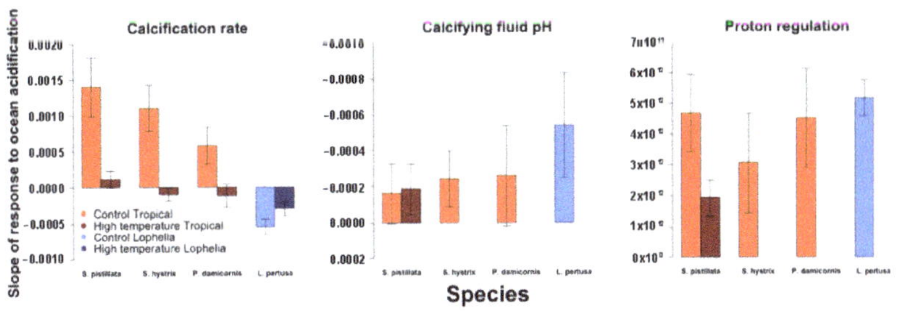

Figure 5. Slopes of regressions calculated from linear mixed effects models investigating the impacts of pCO_2 and temperature on calcification rate, calcifying fluid pH, and $\Delta[H^+]$. Significant differences were found amongst the slopes of the different coral species' calcification responses to ocean acidification and warming. No significant difference was observed amongst the slopes of the different coral species' calcifying fluid pH response to ocean acidification. The slopes of the different coral species' proton regulation response (i.e., $\Delta[H^+]$) to ocean acidification were not significantly different from each other. Vertical bars represent 95% confidence intervals.

3.3. Estimated Coral Photosymbiont Index

Both *S. pistillata* and *P. damicornis* exhibited a lower photosymbiont index (i.e., lower estimated photosymbiont abundance) under the elevated temperature treatments (Figure 6, Table S7). Photosymbiont index significantly increased in both *S. pistillata* and *P. damicornis* when pCO_2 was elevated from control conditions (Figure 6, Table S7). The photosymbiont index of *S. hystrix* was significantly negatively correlated with the interaction between temperature and pCO_2, whereby photosymbiont index increased significantly with increasing pCO_2 under the control temperature treatment, but showed no change in response to pCO_2 in the elevated temperature treatment.

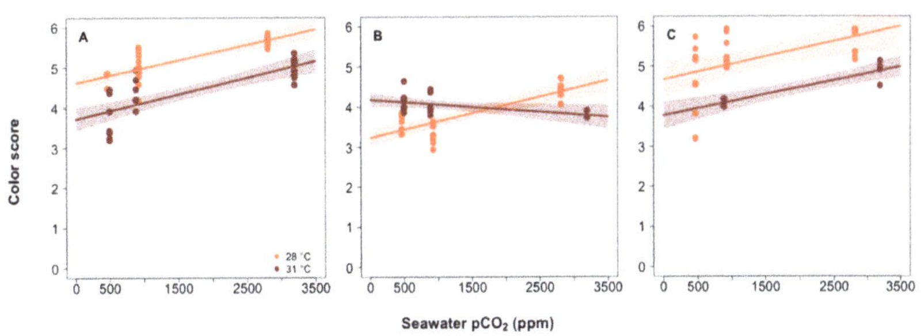

Figure 6. The effect of seawater pCO_2 and warming on the color score of the three species of tropical corals ((**A**): *S. pistillata*; (**B**): *S. hystrix*; (**C**): *P. damicornis*)). Color score is a proxy for photosymbiont abundance ('bleaching'), where 0 = bleached and 6 = healthy. Trendlines indicate significant correlations between seawater pCO_2 and color score at 28 (orange) and 31 (red) °C. The color score of both *S. pistillata* and *P. damicornis* increased in response to increasing pCO_2 (lmers, *S. pistillata*, $p < 0.001$; *P. damicornis*, $p = 0.001$) and decreased in response to a 3 °C temperature increase (lmers, *S. pistillata*, $p < 0.001$, *P. damicornis*, $p = 0.001$). Color score of *S. hystrix* was significantly impacted (indicating bleaching) by an interaction between pCO_2 and temperature (lmer, $p < 0.001$), whereby color score increased significantly with increasing pCO_2 at 28 °C (lmer, $p < 0.001$) but showed no significant change with increasing pCO_2 at 31 °C (lmer, $p = 0.056$). Shaded boundaries represent 95% confidence intervals.

3.4. Investigating the Role of Photosynthesis in Calcification

The calcification rate of *S. pistillata* was best predicted by the significant interaction between photosymbiont index and temperature, whereby photosymbiont index was positively correlated with calcification rate under both temperatures, but the slope of this correlation was significantly greater under the control temperature treatment (Figure 7A, Table S8). The calcification rate of *S. hystrix* was best predicted by a model that included both photosymbiont index and temperature independently. Photosymbiont index had a positive linear relationship with calcification rate under both temperatures, but the slope of this relationship significantly decreased in the high temperature treatment (Figure 7B, Table S8). The calcification rate of *P. damicornis* was best predicted by a model that contained photosymbiont index, seawater pCO_2, and temperature (Table S8). The only significant predictor of calcification rate of *P. damicornis* was the interaction between photosymbiont index, temperature, and pCO_2 (Figure 7C, Table S8).

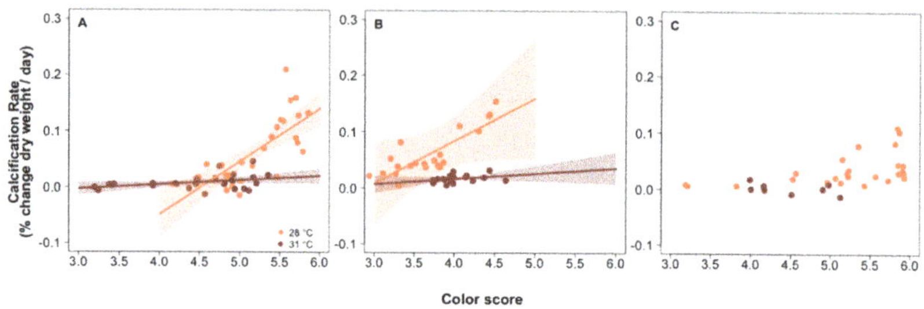

Figure 7. The relationship between coral color score and calcification rate of three species of tropical corals ((**A**): *S. pistillata*; (**B**): *S. hystrix*; (**C**): *P. damicornis*). Color score is a proxy for photosymbiont abundance ('bleaching'), where '0' = bleached and '6' = healthy. Trendlines indicate significant correlations between color score and calcification rate at 28 (orange) and 31 (red) °C. Temperature significantly impacted the relationship between color score and calcification rate of *S. pistillata*, and caused a significant decline in calcification rate of *S. hystrix*. No significant relationship existed between calcification rate and color score of *P. damicornis*, and temperature had no significant impact on this relationship. Shaded boundaries represent 95% confidence intervals.

4. Discussion

4.1. Calcification Response to pCO_2 and Thermal Stress

Prior studies have shown that *S. pistillata* and *P. damicornis* both exhibit resilience in their calcification response to ocean acidification [13,51,52], while *S. hystrix* [53] tends to exhibit more negative responses. The present study, however, shows that, in the absence of thermal stress, these three common species of tropical zooxanthellate corals exhibit increased rates of calcification in response to a one-month exposure to CO_2-acidified conditions. Although the disparity between the results of the past and present studies on these species could be due to differences in experimental design, such as experimental duration, the method used to measure calcification rates, temperature treatments, and/or light levels, these findings provide compelling evidence that, under certain circumstances (e.g., absence of thermal stress), some tropical zooxanthellate coral species can tolerate OA over at least one-month intervals.

Under the elevated temperature treatment (31 °C), the calcification rates of all three tropical coral species were reduced compared to the control temperature, but were unchanged by increasing pCO_2, showing that thermal stress effectively impaired the zooxanthellate corals' calcification response to CO_2-induced OA. However, it should be noted that warming can have either positive or negative effects on coral calcification rate, depending on whether the warming causes temperatures to approach or exceed, respectively, the coral's thermal optimum [19,21,54,55].

Few studies have investigated the calcification response of corals to combined ocean acidification and warming. While some of the studies show a negative response to these combined stressors [55], others contrast the results of the present study by showing no interactive effects of ocean acidification and warming [21,56,57]. The results presented here show that the impacts of ocean acidification on the calcification rates of three *Pocilloporid* coral species are exacerbated by warming. Because OA and global warming typically occur in tandem during major perturbations to the Earth's carbon cycle, both throughout Earth history [58] and as a consequence of anthropogenic CO_2 emissions [59], future CO_2-induced global change poses a substantial threat to these coral species.

The calcification rate of the cold-water azooxanthellate coral *L. pertusa* declined under elevated pCO_2 at both temperatures. This negative calcification response to pCO_2 was weaker under the elevated temperature. Notably, the direction of the pCO_2-temperature interaction for the cold-water azooxanthellate species was opposite that of the three tropical species. *Lophelia pertusa* exhibited net dissolution when seawater pCO_2 reached ca. 1000 ppm, although low levels of calcification have been previously observed for *L. pertusa* under similar conditions [34,35,60]. These conditions are predicted to occur in the surface open ocean by the end of the 21st century [1] and earlier in high-latitude cold-water environments [61]—suggesting that this cold-water ecosystem engineer, whose reef-systems function as nursery ground for a number of commercially important fish species [62], may be unable to form reefs beyond this century. However, the observation that increased temperature partially mitigates the impact of OA agrees with the results of longer-term studies [35], and suggests that the impacts of global change on this species will vary with temperature across latitude and depth, as this species can inhabit seawater ranging from 4 to 14 °C [63,64].

The differences in calcification response to ocean acidification shown here may arise from differences in the physiology and ecology of tropical vs. cold-water corals. Although tropical corals exist close to their thermal limits and are therefore vulnerable to even small amounts of warming, azooxanthellate cold-water corals can generally tolerate a wider temperature range [65]. Elevated respiration rates have been observed for *L. pertusa* in warmer temperatures [66]. Thus, an increase in temperature may boost metabolic rates to partially mitigate the impacts of OA on calcification in the high pCO_2 treatments. The species of tropical corals studied here are colonial and, thus, share resources between closely packed polyps [67]. These polyps are connected by coenosarc tissue, which covers and protects the skeleton [68]. The high degree of tissue cover exhibited by tropical corals means that the skeleton is well protected from dissolution, which may explain the lack of negative calcification response (i.e., lack of net dissolution) for the tropical corals in this study. Their shared gastrovascular system allows the distribution of metabolites generated from coral respiration and zooxanthellate photosynthesis across the colony, which could yield further resilience against ocean acidification. In contrast, *L. pertusa* is a pseudocolonial species [69] that produces single polyps on the end of stalk-like branches. The coenosarc connecting these branches is often partially absent in laboratory specimens and in wild specimens during the winter, leaving regions of exposed skeleton vulnerable to dissolution [65]. This lack of connectivity between polyps and the presence of exposed skeleton may increase the vulnerability of *L. pertusa* to ocean acidification.

Alternatively, the increased solubility of CO_2 in colder waters caused Ω_A of the *L. pertusa* treatments to be 0.12–0.70 units lower at 9 °C than at 12 °C for a given pCO_2 condition. This may have caused higher rates of skeletal dissolution in the high-pCO_2 treatments maintained at the lower temperature, although the rate of dissolution of coral skeletons should be higher under the higher temperature treatments for equivalent Ω_A [3]. Since the buoyant weight method [70] used here yields only a net rate of calcification, i.e., mass of new skeleton produced through gross calcification minus mass of exposed skeleton lost through gross dissolution, it is not possible to determine whether the positive impact of the interaction between pCO_2 and temperature on *L. pertusa* calcification rate was driven by increased gross calcification or reduced gross dissolution (or a combination of these factors) in the high-temperature, high-pCO_2 treatments. Additionally, it was not determined whether OA impacted the density of coral skeleton, which could increase the fragility of the reef framework that these corals form [71].

The responses observed in the present study are consistent with other laboratory studies showing that scleractinian corals exhibit a wide range of calcification responses to OA [13,19–24,52,72]. Some of this variability may arise from differences in experimental design, such as the amount of food provided to the corals, the levels of irradiance, and the duration of the experiment. Nevertheless, this high variability in calcification response patterns across and within species indicates that a greater understanding of the mechanisms that drive coral responses to OA is needed.

The present study was conducted over an eight-week interval, with a total of three weeks of acclimation to laboratory and experimental conditions, and should, therefore, be considered intermediate in duration. As with any experimental OA study, it is possible that the duration of exposure may influence results, as corals may function normally over short timeframes, but exhibit impaired function over longer timeframes as a result of cumulative stress and/or depletion of metabolic resources [19]. Alternatively, corals may exhibit impaired responses shortly after exposure to the treatment conditions due to shock, but acclimate to the treatment conditions over longer timeframes. Prior laboratory-based studies have shown that tropical corals exhibit variable degrees of acclimation to OA over relatively short timescales [19,24,73], whereas acclimation of *L. pertusa* has been observed over longer timescales [74]. The large inter- and intra-specific differences in coral response to OA across experiments and timeframes underscores the need for additional research into the long-term effects of OA on coral calcification.

4.2. Role of Calcifying Fluid pH Regulation in Coral Response to pCO_2 and Thermal Stress

Coral calcifying fluid pH elevation has been widely cited as a mechanism for promoting calcification under conditions of ocean acidification [8,11,14]. Although pH_{CF} declined in all four species under elevated pCO_2, it always remained higher than pH_{SW}. Notably, coral species that showed the highest degree of control over pH_{CF} in the present experiment, and thus the shallowest slope of change in pH_{CF} in response to changing seawater pCO_2, also exhibited the greatest increase in calcification rate when pCO_2 was increased, suggesting that pH_{CF} regulation confers resilience to corals exposed to OA.

These trends are consistent with prior estimation of coral pH_{CF} from boron isotopes [10,38,39,41,75–77], pH-sensitive fluorescent dyes [12], and pH-sensitive microsensors [9,11]. Previous studies of the effects of ocean acidification on calcification site pH show that both *S. pistillata* and *P. damicornis* elevate pH_{CF} above seawater pH, and that this elevation increases under ocean acidification [13,52]. These results are consistent with the findings here, although the measured pH_{CF} was considerably higher in the present study compared to previous studies.

The present study used pH-sensitive microelectrodes, whereas prior studies on both *S. pistillata* and *P. damicornis* [52], and *P. damicornis* [13] used confocal microscopy to image pH-sensitive SNARF-1 dye in coral microcolonies grown on glass slides and boron isotope systematics, respectively. Differences in pH_{CF} could be due to differences in methods of culturing and/or pH_{CF}-estimation, or due to genotypic differences between cultured specimens. Of the three methods used to estimate pH_{CF} (pH-sensitive dyes, boron isotopes, and pH-sensitive microelectrodes), pH-sensitive microelectrodes typically yield the highest pH_{CF} [11], although a side-by-side comparison of pH_{CF} measured with pH-sensitive dye and pH microelectrodes on the same specimens yielded comparable results [14].

Measurements of pH_{CF} using pH-sensitive microelectrodes are challenged by the difficulty in assessing the precise location of the microelectrode tip relative to the coral's calcifying fluid. This challenge was addressed in the present experiment through the creation of pH_{CF} profiles as the pH microelectrode was withdrawn from the calcifying fluid, thus allowing characterization of the calcifying fluid pH compared to intratissue and/or gastrovascular pH (Figure 2), as was conducted in previous studies [14,78]. Additionally, the pH_{CF} of two coral species in the present study (*S. pistillata* and *P. damicornis*) was estimated by coral skeletal $\delta^{11}B$ to provide a side-by-side comparison of these independent approaches to measuring pH_{CF} [41]. A significant correlation was found between pH_{CF}

measured using these two methods, increasing confidence that the measurements obtained here represent pH of the calcifying fluid. The offset between the two approaches was attributed to the two techniques measuring pH_{CF} over different timescales—with skeletal $\delta^{11}B$ recording a time-averaged value of pH_{CF} and pH microelectrodes recording a more instantaneous value of pH_{CF} [41].

Although numerous studies have shown that OA reduces coral pH_{CF}, few have investigated the combined impact of warming and OA on pH_{CF}. In the present study, microelectrode measurements of the pH_{CF} of *S. pistillata* were measured in all pCO_2 and temperature treatments. The prescribed temperature increase resulted in a significant decline in pH_{CF} for each of the three pCO_2 treatments. Under heat stress, corals may receive less nourishment from their photosymbionts to support pH_{CF} regulation due to thermally induced bleaching and/or may divert energy from the regulation of pH_{CF} towards tissue repair.

Assuming that the coral calcifying fluid is ultimately derived from the coral's surrounding seawater [79], the extent to which a coral mitigates the impacts of OA by removing protons from its calcifying fluid can be grossly estimated (excluding the effects of buffering) from the difference between the $[H^+]$ of the calcifying fluid and the $[H^+]$ of the surrounding seawater (i.e., $\Delta[H^+]$). All four coral species exhibited increased $\Delta[H^+]$ under elevated pCO_2 treatments, suggesting that more energy is allocated to pH_{CF} regulation under elevated pCO_2. The $\Delta[H^+]$ was lower in the high-temperature treatment in *S. pistillata*, suggesting that less energy is available to maintain elevated pH_{CF} when the corals experience thermally induced symbiont loss—potentially due to a commensurate decrease in photosynthate translocated to the coral host.

4.3. Role of Photosymbionts in Coral Response to pCO_2 and Thermal Stress

In order for corals to allocate more energy toward removing protons from their calcifying fluid under OA conditions, they must divert energy from other activities and/or increase their energetic intake [80]. Zooxanthellate corals obtain energy from two sources—heterotrophic feeding and photosynthate translocated from their algal symbionts. As corals are sessile suspension feeders, they are limited in the extent to which they can increase heterotrophic feeding, although they may increase feeding rates if sufficient food is available [81,82]. Alternatively, enhanced photosynthesis under OA may play an important role in driving proton elevation under OA.

Using the well-established colorimetric method of Siebeck et al. [45] to estimate the abundance of zooxanthellae via the photosymbiont index, it was shown that the populations of photosymbionts from *S. pistillata* and *P. damicornis* were significantly reduced in the high temperature treatment compared to the initial values. This is consistent with prior work showing that prolonged heat stress can lead to expulsion of zooxanthellate and tissue damage, a process termed 'bleaching' [83]. However, in this study, the photosymbiont index increased in response to increasing pCO_2 under both temperature treatments, suggesting that the photosymbionts within all three *Pocilloporid* tropical corals benefited from the increased availability of dissolved inorganic carbon (DIC). The alleviation of carbon limited photosynthesis could free up energy for elevating pH_{CF} and/or the production of carbon concentrating enzymes (e.g., carbonic anhydrase [84]), thereby aiding calcification under the control temperature [68,85,86]. The zooxanthellae's apparently positive response to increasing DIC also suggests that dissolved inorganic nutrients (DIN) were sufficient and in the correct balance (i.e., Redfield ratio) to sustain photosynthesis [87,88]. This link between enhanced photosynthesis and enhanced calcification under the elevated pCO_2 and control temperature treatment is consistent with the observed correlations between photosymbiont index and calcification rate in *S. pistillata* and *S. hystrix* across all treatments.

4.4. Proposed Mechanistic Framework for Zooxanthellate Coral Response to pCO_2 and Thermal Stress

We propose the following mechanistic framework to explain the zooxanthellate coral responses to OA and warming observed in this study. Under the control temperature and pCO_2 conditions, zooxanthellae fix DIC as carbohydrates (photosynthate), which is then used by the coral hosts as an energy source for all physiological activities, including elevation of pH_{CF} in support of calcification [4,8,89,90]. When OA occurs without thermal stress, high pCO_2 enhances photosynthesis in the coral species investigated and increases their photosymbiont index. Under conditions of elevated pCO_2, enhanced photosymbiont productivity (evidenced by increased photosymbiont index in this study), will result in a greater abundance of byproducts that may be used by the coral host to increase proton removal from the calcifying fluid (evidenced by elevated $\Delta[H^+]$). The combination of elevated pH_{CF} and elevated DIC (due to elevated pCO_2) under OA conditions may allow some species of corals to maintain an Ω_A in their calcifying fluid that is comparable to or, perhaps, greater than those exhibited under non-acidified conditions [11]—hence, their observed ability to maintain constant or, in some cases, elevated rates of calcification under OA.

Although CO_2-induced OA appears supportive of calcification for these three species under the control temperature, this support breaks down in the high temperature treatment. The thermal stress induced in this treatment caused a reduction in the abundance of the corals' algal symbionts (evidenced by their reduced photosymbiont index), which was accompanied by a reduction in $\Delta[H^+]$ (i.e., proton removal) and calcification rate of *S. pistillata* under each of the elevated pCO_2 treatments. It appears that the thermally induced reduction in photosymbiont index eliminated the benefit of enhanced photosynthesis under conditions of elevated pCO_2, thereby leaving the coral with fewer resources (e.g., translocated photosynthate) for elevating pH_{CF}. It should also be noted that thermal impairment of the enzymes used to remove protons from the coral calcifying fluid (e.g., H^+/Ca^{2+}-ATPase [9,91]) may have contributed—along with reduced photosymbiont index—to declines in pH_{CF} and $\Delta[H^+]$ observed for *S. pistillata* in response to thermal stress. Additionally, the different responses shown in the control and high temperature treatments could be due to a temperature-induced shift in the strategy used by the coral to elevate aragonite saturation state in their calcifying fluid. In natural reef systems, it has been shown that coral pH_{CF} is most elevated in the winter months [16]. Although pH_{CF} is still elevated in the warmer summer months, it is elevated to a lesser extent, and DIC elevation appears to be the primary means of raising Ω_A [16]. These observations are consistent with and provide a potential explanation for the results of the present study.

The role of symbiotic zooxanthellae in conferring resilience to corals exposed to OA is also highlighted in the stark contrast observed between the tropical and deep-sea coral responses to OA. Cold-water corals are azooxanthellate and thus do not receive the benefits of enhanced symbiont photosynthesis under elevated pCO_2. Although $\Delta[H^+]$ of the cold-water species was comparable to that of the tropical species under OA, proton regulation of the calcifying fluid probably consumes a greater proportion of the total resources of the cold-water species (which acquires no resources from photosynthesis) than that of the tropical zooxanthellate species. This may leave proportionally fewer resources (compared with the tropical species) for other processes associated with calcification in the cold-water species, such as the production of organic matrices that may initiate crystal nucleation [92], which may explain the more negative calcification response to OA exhibited by *L. pertusa*.

4.5. Limitations of Laboratory-Based Experiments

The results described here were obtained in a controlled laboratory setting. In their natural reef environments, tropical corals experience fluctuations in temperature and carbonate chemistry across daily [93–95] and seasonal [96,97] cycles, whereas corals inhabiting deeper, colder waters experience more stable environments. Thus, differing degrees of prior exposure to fluctuations in pH could contribute to the differential responses of the tropical and cold water coral species observed here. The enhanced decline in pH_{CF} under elevated temperature exhibited by the tropical corals could also result from a shift in strategy for elevating calcifying fluid Ω_A from pH_{CF} elevation to DIC elevation, as has been shown to occur seasonally in a natural reef system [16]. Future research is necessary to assess whether the responses observed here hold in more dynamic temperature and pH environments.

5. Conclusions

Global warming is considered to be amongst the greatest threats facing coral reefs, and OA is emerging as an equally grave threat. We show that warming has a more negative impact than OA on three species of zooxanthellate tropical corals, whereas OA has a more negative impact than warming on an azooxanthellate cold-water coral. This study also provides insight into the role of photosymbionts in corals' response to OA. Specifically, the enhancement of symbiont photosynthesis under higher-pCO_2 conditions appears to mitigate the negative effects of OA on tropical zooxanthellate corals by providing resources that assist in the maintenance of elevated calcifying fluid pH in support of calcification. This resilience, however, is impaired when OA is combined with thermally induced reductions in the abundance of the coral's photosymbionts (i.e., 'bleaching'), which limits the extent to which the coral holobiont can utilize the elevated DIC via photosynthesis. These results highlight the threat that ocean warming and acidification pose for tropical and cold-water corals, especially when occurring in tandem.

Supplementary Materials: The following supporting information can be downloaded at: https://www.mdpi.com/article/10.3390/jmse10081106/s1: Summary of experimental design and results (Section S1); Water quality methods and summary tables (Section S2); Buoyant weight methods (Section S3); Images of calcein dye incorporation into coral skeleton (Section S4); Method for estimating photosymbiont index (Section S5); and Model summary tables (Section S6).

Author Contributions: Conceptualization, J.B.R. and R.A.E.; methodology, J.B.R., R.A.E., L.P.C., D.D.B., H.W., J.V.B., G.M.S.-G. and I.W.; formal analysis, L.P.C., J.B.R. and J.G.; investigation, L.P.C., J.B.R., C.E.R., F.M.-L., I.W., J.V.B. and R.A.E. sources, J.B.R., R.A.E., D.D.B., J.B. and H.W.; data curation, L.P.C., J.B.R. and R.A.E.; writing—original draft preparation, L.P.C. and J.B.R.; writing—review and editing, L.P.C., J.B.R., R.A.E., M.G., C.E.R., J.B., J.V.B., D.D.B., G.M.S.-G. and H.W.; visualization, L.P.C., J.B.R. and J.G.; supervision, J.B.R., R.A.E., D.D.B., J.B. and H.W.; project administration, J.B.R. and H.W.; funding acquisition, J.B.R., R.A.E. and H.W. All authors have read and agreed to the published version of the manuscript.

Funding: J.B.R. acknowledges support from National Science Foundation grant OCE-1437371, the ZMT, and a Hanse-Wissenschaftskolleg Fellowship. R.A.E. acknowledges support from National Science Foundation grant OCE-1437166, the Pritzker Endowment to UCLA IoES, and 'Laboratoire d'Excellence' LabexMER grant ANR-10-LABX-19 co-funded by a grant from the French government under the program 'Investissements d'Avenir'.

Institutional Review Board Statement: Not applicable.

Informed Consent Statement: Not applicable.

Data Availability Statement: It is our intention to make data available on publication of this study.

Acknowledgments: We thank Artur Fink, Laurie Hoffman, and Anja Niclas (MPIMM) for assistance with construction and use of pH microelectrodes; Silvia Hardenberg, Nico Steinel, and Christian Brandt (ZMT) for assistance with coral husbandry and redesign of the acidification system at the ZMT; and Matthias Birkicht (ZMT) for assistance with water chemistry analysis.

Conflicts of Interest: The authors declare no conflict of interest.

References

1. IPCC. *Climate Change 2013—The Physical Science Basis*; Intergovernmental Panel on Climate Change, Ed.; Cambridge University Press: Cambridge, UK, 2014; ISBN 9781107415324. [CrossRef]
2. Caldeira, K.; Wickett, M.E. Anthropogenic carbon and ocean pH. *Nature* **2003**, *425*, 365. [CrossRef] [PubMed]

3. Ries, J.B.; Ghazaleh, M.N.; Connolly, B.; Westfield, I.; Castillo, K.D. Impacts of seawater saturation state state (Ω_A = 0.4–4.6) and temperature (10, 25 °C) on the dissolution kinetics of whole-shell biogenic carbonates. *Geochim. Cosmochim. Acta* **2016**, *192*, 318–337. [CrossRef]
4. Jokiel, P.L.; Rodgers, K.S.; Kuffner, I.B.; Andersson, A.J.; Cox, E.F.; Mackenzie, F.T. Ocean acidification and calcifying reef organisms: A mesocosm investigation. *Coral Reefs* **2008**, *27*, 473–483. [CrossRef]
5. Milliman, J.D. Production and accumulation of calcium carbonate in the ocean: Budget of a nonsteady state. *Glob. Biogeochem. Cycles* **1993**, *7*, 927–957. [CrossRef]
6. Anthony, K.R.; Fabricius, K.E. Shifting roles of heterotrophy and autotrophy in coral energetics under varying turbidity. *J. Exp. Mar. Bio. Ecol.* **2000**, *252*, 221–253. [CrossRef]
7. Hoegh-Guldberg, O. Climate change, coral bleaching and the future of the world's coral reefs. *Mar. Freshw. Res.* **1999**, *50*, 839–866. [CrossRef]
8. McCulloch, M.; Falter, J.; Trotter, J.; Montagna, P. Coral resilience to ocean acidification and global warming through pH up-regulation. *Nat. Clim. Change* **2012**, *2*, 623–627. [CrossRef]
9. Al-Horani, F.A.; Al-Moghrabi, S.M.; de Beer, D. The mechanism of calcification and its relation to photosynthesis and respiration in the scleractinian coral *Galaxea fascicularis*. *Mar. Biol.* **2003**, *142*, 419–426. [CrossRef]
10. Allison, N.; Finch, A.A. 11B, Sr, Mg and B in a modern *Porites* coral: The relationship between calcification site pH and skeletal chemistry. *Geochim. Cosmochim. Acta* **2010**, *74*, 1790–1800. [CrossRef]
11. Ries, J.B. A physicochemical framework for interpreting the biological calcification response to CO_2-induced ocean acidification. *Geochim. Cosmochim. Acta* **2011**, *75*, 4053–4064. [CrossRef]
12. Venn, A.; Tambutté, E.; Holcomb, M.; Allemand, D.; Tambutté, S. Live tissue imaging shows reef corals elevate pH under their calcifying tissue relative to seawater. *PLoS ONE* **2011**, *6*, e20013. [CrossRef] [PubMed]
13. Comeau, S.; Cornwall, C.E.; McCulloch, M.T. Decoupling between the response of coral calcifying fluid pH and calcification to ocean acidification. *Sci. Rep.* **2017**, *7*, 7573. [CrossRef] [PubMed]
14. Sevilgen, D.S.; Venn, A.A.; Hu, M.Y.; Tambutté, E.; de Beer, D.; Planas-Bielsa, V.; Tambutté, S. Full in vivo characterization of carbonate chemistry at the site of calcification in corals. *Sci. Adv.* **2019**, *5*, eaau7447. [CrossRef] [PubMed]
15. Allison, N.; Cohen, I.; Finch, A.A.; Erez, J.; Tudhope, A.W. Corals concentrate dissolved inorganic carbon to facilitate calcification. *Nat. Commun.* **2014**, *5*, 5741. [CrossRef] [PubMed]
16. McCulloch, M.; D'Olivo, J.P.; Falter, J.; Holcomb, M.; Trotter, J.A. Coral calcification in a changing world and the interactive dynamics of pH and DIC upregulation. *Nat. Commun.* **2017**, *8*, 15686. [CrossRef]
17. Ries, J.B. Skeletal mineralogy in a high-CO_2 world. *J. Exp. Mar. Biol. Ecol.* **2011**, *403*, 54–64. [CrossRef]
18. Spalding, C.; Finnegan, S.; Fischer, W.W. Energetic costs of calcification under ocean acidification. *Glob. Biogeochem. Cycles* **2017**, *31*, 866–877. [CrossRef]
19. Castillo, K.D.; Ries, J.B.; Bruno, J.F.; Westfield, I.T. The reef-building coral *Siderastrea siderea* exhibits parabolic responses to ocean acidification and warming. *Proc. Biol. Sci.* **2014**, *281*, 20141856. [CrossRef]
20. Marubini, F.; Ferrier-Pages, C.; Furla, P.; Allemand, D. Coral calcification responds to seawater acidification: A working hypothesis towards a physiological mechanism. *Coral Reefs* **2008**, *27*, 491–499. [CrossRef]
21. Bove, C.B.; Ries, J.B.; Davies, S.W.; Westfield, I.T.; Umbanhowar, J.; Castillo, K.D. Common Caribbean corals exhibit highly variable responses to future acidification and warming. *Proc. Biol. Sci.* **2019**, *286*, 20182840. [CrossRef]
22. Ries, J.B.; Cohen, A.L.; McCorkle, D.C. A nonlinear calcification response to CO_2-induced ocean acidification by the coral *Oculina arbuscula*. *Coral Reefs* **2010**, *29*, 661–674. [CrossRef]
23. Reynaud, S.; Leclercq, N.; Romaine-Lioud, S.; Ferrier-Pagès, C.; Jaubert, J.; Gattuso, J.-P. Interacting effects of CO_2 partial pressure and temperature on photosynthesis and calcification in a scleractinian coral. *Glob. Chang. Biol.* **2003**, *9*, 1660–1668. [CrossRef]
24. Comeau, S.; Cornwall, C.E.; DeCarlo, T.M.; Doo, S.S.; Carpenter, R.C.; McCulloch, M.T. Resistance to ocean acidification in coral reef taxa is not gained by acclimatization. *Nat. Clim. Chang.* **2019**, *9*, 477. [CrossRef]
25. Carafoli, E. Calcium signaling: A tale for all seasons. *Proc. Natl. Acad. Sci. USA* **2002**, *99*, 1115–1122. [CrossRef]
26. Mass, T.; Drake, J.L.; Peters, E.C.; Jiang, W.; Falkowski, P.G. Immunolocalization of skeletal matrix proteins in tissue and mineral of the coral *Stylophora pistillata*. *Proc. Natl. Acad. Sci. USA* **2014**, *111*, 12728–12733. [CrossRef]
27. Falini, G.; Fermani, S.; Goffredo, S. Coral biomineralization: A focus on intra-skeletal organic matrix and calcification. *Semin. Cell Dev. Biol.* **2015**, *35*, 17–26. [CrossRef]
28. Marin, F.; Smith, M.; Isa, Y.; Muyzer, G.; Westbroek, P. Skeletal matrices, muci, and the origin of invertebrate calcification. *Proc. Natl. Acad. Sci. USA* **1996**, *93*, 1554–1559. [CrossRef]
29. Westbroek, P.; Marin, F. A marriage of bone and nacre. *Nature* **1998**, *392*, 861–862. [CrossRef]
30. Mass, T.; Drake, J.L.; Haramaty, L.; Kim, J.D.; Zelzion, E.; Bhattacharya, D.; Falkowski, P.G. Cloning and characterization of four novel coral acid-rich proteins that precipitate carbonates in vitro. *Curr. Biol.* **2003**, *23*, 1126–1131. [CrossRef]
31. Mass, T.; Giuffre, A.J.; Sun, C.-Y.; Stifler, C.A.; Frazier, M.J.; Neder, M.; Tamura, N.; Stan, C.V.; Marcus, M.A.; Gilberg, P.U.P.A. Amorphous calcium carbonate particles form coral skeletons. *Proc. Natl. Acad. Sci. USA* **2017**, *114*, E7670–E7678. [CrossRef]
32. Hohn, S.; Reymond, C.E. Coral calcification, mucus, and the origin of skeletal organic molecules. *Coral Reefs* **2019**, *38*, 973–984. [CrossRef]

33. Roberts, J.M.; Wheeler, A.; Freiwald, A.; Cairns, S. *The Biology and Geology of Deep-Sea Coral Habitats*; Cambridge University Press: Cambridge, UK, 2009.
34. Hennige, S.J.; Wicks, L.C.; Kamenos, N.A.; Perna, G.; Findlay, H.S.; Roberts, J.M. Hidden impacts of ocean acidification to live and dead coral framework. *Proc. R. Soc. B* **2015**, *282*, 20150990. [CrossRef] [PubMed]
35. Büscher, J.V.; Form, A.U.; Riebesell, U. Interactive effects of ocean acidification and warming on growth, fitness and survival of the cold-water coral *Lophelia pertusa* under different food availabilities. *Front. Mar. Sci.* **2017**, *4*, 101. [CrossRef]
36. Georgian, S.E.; Dupont, S.; Kurman, M.; Butler, A.; Strömberg, S.M.; Larsson, A.I.; Cordes, E.E. Biogeographic variability in the physiological response of the cold-water coral *Lophelia pertusa* to ocean acidification. *Mar. Ecol.* **2016**, *37*, 1345–1359. [CrossRef]
37. Naumann, M.S.; Orejas, C.; Ferrier-Pagès, C. Species-specific physiological response by the cold-water corals *Lophelia pertusa* and *Madrepora oculata* to variations within their natural temperature range. *Deep. Sea Res. Part II Top. Stud. Oceanogr.* **2014**, *99*, 36–41. [CrossRef]
38. Holcomb, M.; Venn, A.A.; Tambutté, E.; Tambutté, S.; Allemand, D.; Trotter, J.; McCulloch, M. Coral calcifying fluid pH dictates response to ocean acidification. *Sci. Rep.* **2014**, *4*, 5207. [CrossRef]
39. Liu, Y.-W.; Sutton, J.N.; Ries, J.B.; Eagle, R.A. Regulation of calcification site pH is a polyphyletic but not always governing response to ocean acidification. *Sci. Adv.* **2020**, *6*, eaax1314. [CrossRef]
40. D'Olivo, J.P.; McCulloch, M.T. Response of coral calcification and calcifying fluid composition to thermally induced bleaching stress. *Nat. Sci. Rep.* **2017**, *7*, 2207. [CrossRef]
41. Guillermic, M.; Cameron, L.P.; De Corte, I.; Misra, S.; Bijma, J.; de Beer, D.; Reymond, C.E.; Westphal, H.; Ries, J.B.; Eagle, R.A. Thermal stress reduces pocilloporid coral resilience to ocean acidification by impairing control over calcifying fluid chemistry. *Sci. Adv.* **2021**, *7*, eaba9958. [CrossRef]
42. Ellison, J.C.; Fiu, M. Vulnerability of Fiji's mangroves and associated coral reefs to climate change. *Ed. WWF S. Pac. Programme* **2010**, 50p.
43. Cameron, L.P.; Reymond, C.E.; Müller-Lundin, F.; Westfield, I.; Grabowski, J.H.; Westphal, H.; Ries, J.B. Effects of temperature and ocean acidification on the extrapallial fluid pH, calcification rate, and condition factor of the king scallop *Pecten maximus*. *J. Shellfish. Res.* **2019**, *38*, 763–777. [CrossRef]
44. Pierrot, D.; Lewis, E.; Wallace, D.W.R. MS excel program developed for CO_2 system calculations. In *ORNL/CDIAC-105a. Carbon Dioxide Information Analysis Center*; Oak Ridge National Laboratory U.S. Department Energy: Oak Ridge, TN, USA, 2006.
45. Siebeck, U.E.; Marshall, N.J.; Klüter, A.; Hoegh-Guldberg, O. Monitoring coral bleaching using a colour reference card. *Coral Reefs* **2006**, *25*, 453–460. [CrossRef]
46. Conti-Jerpe, I.E.; Thompson, P.D.; Wai Martin Wong, C.; Oliveira, N.L.; Duprey, N.N.; Moynihan, M.A.; Baker, D.M. Trophic strategy and bleaching resistance in reef-building corals. *Sci. Adv.* **2020**, *6*, eaaz5443. [CrossRef] [PubMed]
47. Morgans, C.A.; Hung, J.Y.; Bourne, D.G.; Quigley, K.M. Symbiodiniaceae probiotics for use in bleaching recovery. *Restor. Ecol.* **2020**, *28*, 282–288. [CrossRef]
48. De Beer, D.E.; Schramm, A.; Santegoeds, C.M.; Kühl, M. A nitrite microsensor for profiling environmental biofilms. *Appl. Environ. Microbiol.* **1997**, *63*, 973–977. [CrossRef]
49. Bates, D.; Mächler, M.; Bolker, B.; Walker, S. Fitting linear mixed-effects models using lme. *J. Stat. Softw.* **2015**, *67*, 1–48. [CrossRef]
50. Akaike, H. Factor analysis and AIC. In *Selected Papers of Hirotugu Akaike. Springer Series in Statistics (Perspectives in Statistics)*; Parzen, E., Tanabe, K., Kitagawa, G., Eds.; Springer: New York, NY, USA, 1987.
51. Houlbrèque, F.; Rodolfo-Metalpa, R.; Jeffree, R.; Oberhänsli, F.; Teyssié, J.-L.; Boisson, F.; Al-Trabeen, K.; Ferrier-Pagès, C. Effects of increased pCO_2 on zinc uptake and calcification in the tropical coral *Stylophora pistillata*. *Coral Reefs* **2012**, *31*, 101–109. [CrossRef]
52. Venn, A.A.; Tambutté, E.; Caminiti-Segonds, N.; Techer, N.; Allemand, D.; Tambutté, S. Effects of light and darkness on pH regulation in three coral species exposed to seawater acidification. *Sci. Rep.* **2019**, *9*, 2201. [CrossRef]
53. Rädecker, N.; Meyer, F.W.; Bednarz, V.N.; Cardini, U.; Wild, C. Ocean acidification rapidly reduces dinitrogen fixation associated with the hermatypic coral *Seriatopora hystrix*. *Mar. Ecol. Prog. Ser.* **2014**, *511*, 297–302. [CrossRef]
54. Cantin, N.E.; Cohen, A.L.; Karnauskas, K.B.; Tarrant, A.M.; McCorkle, D.C. Ocean warming slows coral growth in the central Red Sea. *Science* **2010**, *329*, 322–325. [CrossRef]
55. Horvath, K.M.; Castillo, K.D.; Armstrong, P.; Westfield, I.T.; Courtney, T.; Ries, J.B. Next-century ocean acidification and warming both reduce calcification rate, but only acidification alters skeletal morphology of reef-building coral *Siderastrea siderea*. *Sci. Rep.* **2016**, *6*, 29639. [CrossRef] [PubMed]
56. Schoepf, V.; Grottoli, A.G.; Warner, M.E.; Cai, W.-J.; Melman, T.F.; Hoadley, K.D.; Pettay, D.T.; Hu, X.; Li, Q.; Xu, H.; et al. Coral energy reserves and calcification in a high-CO_2 world at two temperatures. *PLoS ONE* **2013**, *8*, e75049. [CrossRef] [PubMed]
57. Okazaki, R.R.; Towle, E.K.; van Hooidonk, R.; Mor, C.; Winter, R.N.; Piggot, A.M.; Cunning, R.; Baker, A.C.; Klaus, J.S.; Swart, P.K.; et al. Species-specific responses to climate change and community composition determine future calcification rates of Florida Keys reefs. *Glob. Chang. Biol.* **2016**, *23*, 1023–1035. [CrossRef] [PubMed]
58. Hönisch, B.; Ridgwell, A.; Schmidt, D.N.; Gibbs, S.J.; Sluijs, A.; Zeebe, R.; Kump, L.; Martindale, R.C.; Greene, S.E.; Kiessling, W.; et al. The geologic record of ocean acidification. *Science* **2012**, *335*, 1058–1063. [CrossRef] [PubMed]
59. Bijma, J.; Pörtner, H.-O.; Yesson, C.; Rogers, A.D. Climate change and the oceans–What does the future hold? *Mar. Pollut. Bull.* **2013**, *74*, 495–505. [CrossRef]

60. Kurman, M.D.; Gómez, C.E.; Georgian, S.E.; Lunden, J.J.; Cordes, E.E. Intra-specific variation reveals potential adaptation to ocean acidification in a cold-water coral from the Gulf of Mexico. *Front. Mar. Sci.* **2017**, *4*, 111. [CrossRef]
61. Fabry, V.J.; McClintock, J.B.; Mathis, J.T.; Grebmeier, J.M. Ocean acidification at high latitudes: The bellwether. *Oceanography* **2009**, *22*, 160–171. [CrossRef]
62. Costello, M.J.; McCrea, M.; Freiwald, A.; Lundälv, T.; Jonsson, L.; Bett, B.J.; van Weering, T.C.E.; de Haas, H.; Roberts, J.M.; Allen, D. Role of cold-water *Lophelia pertusa* coral reefs as fish habitat in the NE Atlantic. In *Cold-Water Corals and Ecosystems. Erlangen Earth Conference Series*; Freiwald, A., Roberts, J.M., Eds.; Springer: Berlin, Germany, 2005; pp. 771–805.
63. Mortensen, P.B.; Hovland, T.; Fossa, J.H.; Furevik, D.M. Distribution, abundance and size of *Lophelia pertusa* coral reefs in mid-Norway in relation to seabed characteristics. *J. Mar. Biol. Assoc. UK* **2001**, *81*, 581–597. [CrossRef]
64. Freiwald, A. Geobiology of *Lophelia pertusa* (scleractinia) reefs in the North Atlantic. Habilitation Thesis, University of Bremen, Bremen, Germany, 1998.
65. Rogers, A.D. The biology of *Lophelia pertusa* (linneaus 1758) and other deep-water reef-forming corals and impacts from human activities. *Int. Rev. Hydrobiol.* **1999**, *84*, 315–406. [CrossRef]
66. Dodds, L.A.; Roberts, J.M.; Taylor, A.C.; Marubini, F. Metabolic tolerance of the cold-water coral *Lophelia pertusa* (scleractinia) to temperature and dissolved oxygen change. *J. Exp. Mar. Biol. Ecol.* **2007**, *349*, 205–214. [CrossRef]
67. Fine, M.; Oren, U.; Loya, Y. Bleaching effect on regeneration and resource translocation in the coral *Oculina patagonica*. *Mar. Ecol. Prog. Ser.* **2002**, *234*, 119–125. [CrossRef]
68. Tambutté, E.; Allemand, D.; Zoccola, D.; Meibom, A.; Lotto, S.; Caminiti, N.; Tambutté, S. Observations of the tissue-skeleton interface in the scleractinian coral *Stylophora pistillata*. *Coral Reefs* **2007**, *26*, 517–529. [CrossRef]
69. Järnegren, J.; Kutti, T. *Lophelia pertusa* in Norwegian waters. What have we learned since 2008? *NINA Rep.* **2014**, *1028*, 40.
70. Davies, P.S. Short-term growth measurements of corals using an accurate buoyant weighing technique. *Mar. Biol.* **1989**, *101*, 389–395. [CrossRef]
71. Hennige, S.J.; Wolfram, U.; Wickes, L.; Murray, F.; Murray Roberts, J.; Kamenos, N.A.; Schofield, S.; Groetsch, A.; Spiesz, E.M.; Aubin-Tam, M.-E.; et al. Crumbling reefs and cold-water coral habitat loss in a future ocean: Evidence of "Coralporosis" as an indicator of habitat integrity. *Front. Mar. Sci.* **2020**, *7*, 668. [CrossRef]
72. Chan, N.C.S.; Connelly, S.R. Sensitivity of coral calcification to ocean acidification: A meta-analysis. *Glob. Chang. Biol.* **2013**, *19*, 282–290. [CrossRef]
73. Davies, S.W.; Marchetti, A.; Ries, J.B.; Castillo, K.D. Thermal and pCO_2 stress elicit divergent transcriptomic responses in a resilient coral. *Front. Mar. Sci.* **2016**, *3*, 112. [CrossRef]
74. Form, A.U.; Riebesell, U. Acclimation to ocean acidification during long-term CO_2 exposure in the cold-water coral *Lophelia pertusa*. *Glob. Chang. Biol.* **2012**, *18*, 843–853. [CrossRef]
75. Tanaka, K.; Holcomb, M.; Takahashi, A.; Kurihara, H.; Asami, R.; Shinjo, R.; Sowa, K.; Rankenburg, K.; Watanabe, T.; McCulloch, M. Response of *Acropora digitifera* to ocean acidification: Constraints from ^{11}B, Sr, Mg, and Ba compositions of aragonitic skeletons cultured under variable seawater pH. *Coral Reefs* **2015**, *34*, 1139–1149. [CrossRef]
76. Comeau, S.; Cornwall, C.E.; DeCarlo, T.M.; Krieger, E.; McCulloch, M.T. Similar controls on calcification under ocean acidification across unrelated coral reef taxa. *Glob. Chang. Biol.* **2018**, *24*, 4857–4868. [CrossRef]
77. Sutton, J.N.; Liu, Y.-W.; Ries, J.B.; Guillermic, M.; Ponzevera, E.; Eagle, R.A. ^{11}B as monitor of calcification site pH in divergent marine organisms. *Biogeosciences* **2018**, *15*, 1447–1467. [CrossRef]
78. Cai, W.-J.; Ma, Y.; Hopkinson, B.M.; Grottoli, A.G.; Warner, M.E.; Ding, Q.; Hu, X.; Yuan, X.; Schoepf, V.; Xu, H.; et al. Microelectrode characterization of coral daytime interior pH and carbonate chemistry. *Nat. Commun.* **2016**, *7*, 11144. [CrossRef] [PubMed]
79. Cohen, A.L.; McConnaughey, T.A. Geochemical perspectives on coral mineralization. *Rev. Mineral. Geochem.* **2003**, *54*, 151–187. [CrossRef]
80. Kooijman, B. *Dynamic Energy Budget Theory for Metabolic Organization*; Cambridge University Press: Cambridge, UK, 2009.
81. Holcomb, M.; McCorkle, D.C.; Cohen, A.L. Long-term effects of nutrient and CO_2 enrichment on the temperate coral *Astrangia poculata* (Ellis and Solander, 1786). *J. Exp. Mar. Biol. Ecol.* **2010**, *386*, 27–33. [CrossRef]
82. Edmunds, P.J. Zooplanktivory ameliorates the effects of ocean acidification on the reef coral *Porites* spp. *Limnol. Oceanogr.* **2011**, *56*, 2402–2410. [CrossRef]
83. Gates, R.D.; Baghdasarian, G.; Muscatine, L. Temperature stress causes host cell detachment in symbiotic cnidarians: Implications for coral bleaching. *Biol. Bull.* **1992**, *182*, 324–332. [CrossRef]
84. Brading, P.; Warner, M.E.; Davey, P.; Smith, D.J.; Achterberg, E.P.; Suggett, D.J. Differential effects of ocean acidification on growth and photosynthesis among phylotypes of *Symbiodinium* (Dinophyceae). *Limnol. Oceanogr.* **2011**, *56*, 927–938. [CrossRef]
85. Moya, A.; Tambutté, S.; Bertucci, A.; Tambutté, E.; Lotto, S.; Vullo, D.; Supuran, C.T.; Allemand, D.; Zoccola, D. Carbonic anhydrase in the scleractinian coral *Stylophora pistillata* characterization, localization, and role in biomineralization. *J. Biol. Chem.* **2008**, *283*, 25475–25484. [CrossRef]
86. Chen, S.; Gagnon, A.C.; Adkins, J.F. Carbonic anhydrase, coral calcification and a new model of stable isotope vital effects. *Geochim. Cosmochim. Acta* **2018**, *236*, 179–197. [CrossRef]
87. Muscatine, L.; McCloskey, L.R.; Marian, R.E. Estimating the daily contribution of carbon from zooxanthellae to coral animal respiration. *Limnol. Oceanogr.* **1981**, *26*, 601–611. [CrossRef]

88. Dubinsky, Z.; Jokiel, P.L. Ratio of energy and nutrient fluxes regulates symbiosis between zooxanthellae and corals. *Pac. Sci.* **1994**, *48*, 313–324.
89. Aichelman, H.E.; Bove, C.B.; Castillo, K.D.; Boulton, J.M.; Knowlton, A.C.; Nieves, O.C.; Ries, J.B.; Davies, S.W. Exposure duration modulates the response of Caribbean corals to global change stressors. *Limnol. Oceanogr. Lett.* **2021**, *66*, 8. [CrossRef]
90. Cornwall, C.E.; Comeau, S.; DeCarlo, T.M.; Moore, B.; D'alexis, Q.; McCulloch, M.T. Resistance of corals and coralline algae to ocean acidification: Physiological control of calcification under natural pH variability. *Proc. R. Soc. B* **2018**, *285*, 20181168. [CrossRef]
91. Cohen, A.L.; Holcomb, M. Why corals care about ocean acidification: Uncovering the mechanism. *Oceanography* **2009**, *22*, 118–127. [CrossRef]
92. Allemand, D.; Ferrier-Pagès, C.; Furla, P.; Houlbrèque, F.; Puverel, S.; Reynaud, S.; Tambutté, É.; Tambutté, S.; Zoccola, D. Biomineralization in reef-building corals: From molecular mechanisms to environmental control. *Comptes Rendus Palevol* **2004**, *3*, 453–467. [CrossRef]
93. Gagliano, M.; McCormick, M.I.; Moore, J.A.; Depczynski, M. The basics of acidification: Baseline variability of pH on Australian coral reefs. *Mar. Biol.* **2010**, *157*, 1849–1856. [CrossRef]
94. Hofmann, G.E.; Smith, J.E.; Johnson, K.S.; Send, U.; Levin, L.A.; Micheli, F.; Paytan, A.; Price, N.N.; Peterson, B.; Takeshita, Y.; et al. High-frequency dynamics of ocean pH: A multi-ecosystem comparison. *PLoS ONE* **2011**, *6*, e28983. [CrossRef]
95. Cyronak, T.; Takeshita, Y.; Courtney, T.A.; DeCarlo, E.H.; Eyre, B.D.; Kline, D.I.; Martz, T.; Page, H.; Price, N.N.; Smith, J.; et al. Diel temperature and pH variability scale with depth across diverse coral reef habitats. *Limnol. Oceanogr. Lett.* **2019**, *5*, 193–203. [CrossRef]
96. Gray, S.E.C.; DeGrandpre, M.D.; Langdon, C.; Corredor, J.E. Short-term and seasonal pH, pCO_2 and saturation state variability in a coral-reef ecosystem. *Glob. Biogeochem. Cycles* **2012**, *26*, GB3012. [CrossRef]
97. Kline, D.I.; Teneva, L.; Hauri, C.; Schneider, K.; Miard, T.; Chai, A.; Marker, M.; Dunbar, R.; Caldeira, K.; Lazar, B.; et al. Six month in situ high-resolution carbonate chemistry and temperature study on a coral reef flat reveals asynchronous pH and temperature anomalies. *PLoS ONE* **2015**, *10*, e0127648. [CrossRef]

Journal of
Marine Science and Engineering

MDPI

Article

Why Do Bio-Carbonates Exist?

Luis Pomar [1], Pamela Hallock [2,*], Guillem Mateu-Vicens [3] and Juan I. Baceta [4]

1 Catedra Guillem Colom, Universitat de les Illes Balears, E-07122 Palma de Mallorca, Spain
2 College of Marine Science, University of South Florida, St. Petersburg, FL 33701, USA
3 Department of Biology, Universitat de les Illes Balears, E-07122 Palma de Mallorca, Spain
4 Department of Geology, University of the Basque Country UPV-EHU, E-48940 Leioa, Spain
* Correspondence: pmuller@usf.edu

Abstract: Calcium carbonate precipitation associated with biotic activity is first recorded in Archaean rocks. The oldest putative fossils related to hydrothermal vents have been dated at ~3.77 Ga (possibly 4.29 Ga). Stromatolites, the oldest dated at 3.70 Ga, have since occurred through Earth history, despite dramatic changes in physical and chemical conditions in aquatic environments. A key question is: what advantages do photosynthesizing aquatic prokaryotes and algae gain by precipitating carbonates? We propose the Phosphate Extraction Mechanism (PEM) to explain the benefits of biomineralization in warm, oligotrophic, alkaline, euphotic environments. Carbonate precipitation enhances access to otherwise limited carbon dioxide and phosphate in such environments. This mechanism also provides an explanation for prolific production of carbonates during times of elevated atmospheric carbon dioxide at intervals in the Phanerozoic.

Keywords: phosphate; nutrient limitation; carbon dioxide; Archaean; Proterozoic; cyanobacteria; calcareous algae; coccolithophores

Citation: Pomar, L.; Hallock, P.; Mateu-Vicens, G.; Baceta, J.I. Why Do Bio-Carbonates Exist?. *J. Mar. Sci. Eng.* **2022**, *10*, 1648. https://doi.org/10.3390/jmse10111648

Academic Editors: Hildegard Westphal, Justin Ries and Steve Doo

Received: 27 September 2022
Accepted: 21 October 2022
Published: 3 November 2022

1. Introduction

Carbon (C), which is the major component of fossil fuels [1], is the key ingredient in "life" [2]. Carbon is also a key ingredient in limestones ($CaCO_3$) and other carbonate rocks [3]. The ultimate sources of carbon are the primordial components of the Earth [2]. Even at present, volcanic sources emit ~6×10^8 metric tons of carbon dioxide (CO_2) per year [4]. Through geologic history, the dynamics of carbon chemistry have played major roles in oceanic, terrestrial and even subsurface-crustal processes [2]. Human activities are now contributing roughly 40 billion tons of CO_2 annually into the Earth's atmosphere [5], with climatic consequences that are becoming more apparent every year.

The systematic measure of CO_2 concentrations in the atmosphere was started by C. David Keeling of the Scripps Institution of Oceanography in March 1958 at a NOAA (National Oceanic and Atmospheric Administration) facility at Mauna Loa Observatory, Hawai'i, USA [6]. NOAA initiated its own CO_2 measurement in May 1974 and, since then, NOAA and Scripps have run the measurements in parallel [7]. Prior to Keeling's research, CO_2 measurements were inconsistent, but the Keeling's methods revealed a pattern of seasonal oscillations of CO_2, with peaks in May and lows in November, with the averaged values of successive years progressively increasing (Figure 1). Keeling envisaged the yearly cycles to reflect the vegetation cycles that prevail across the northern hemisphere while the increase over time was thought to be caused by human activities, especially the burning of fossil fuels [8].

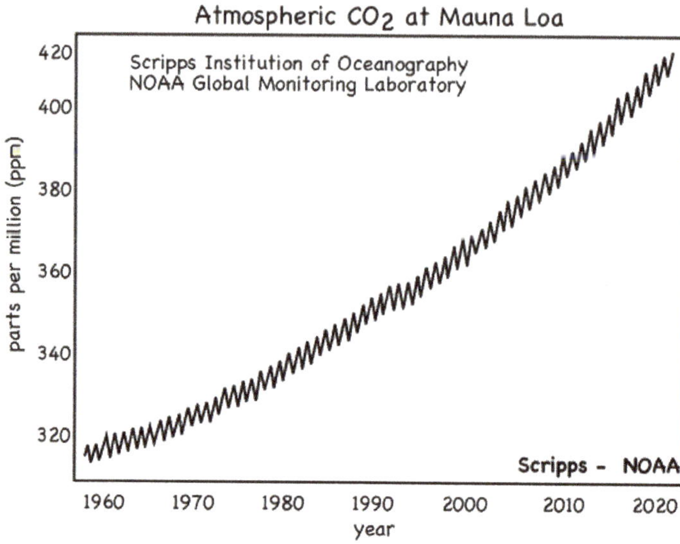

Figure 1. CO_2 concentrations in the atmosphere [7], started by C. David Keeling of the Scripps Institution of Oceanography in March 1958 [8]. (Adapted with permission from Scripps CO_2 program, Scripps Institution of Oceanography at UC San Diego, CA, USA.).

By the 1960′s, greenhouse gas emissions and their link to global climate change became a serious concern, with both scientists and the public considering CO_2 to be an invisible pollutant. Attention and concern for greenhouse gases and their role in climate change intensified, and in 1988 the United Nations created the Intergovernmental Panel on Climate Change (IPCC) [9]. Concern for ocean acidification, *climate change's equally evil twin*, emerged somewhat later. Although scientists had been tracking ocean pH for more than 30 years, biological studies emerged in the 1990s [10] and have accelerated since the introduction of the term "ocean acidification" in 2003 [11].

Burning of fossil fuels, combined with widespread changes in land use, has resulted in rapidly increasing concentrations of atmospheric CO_2 that are causing the decline in the pH of surface seawater [10–14]. Will progressive warming and acidification cause mass extinctions and evolutionary turnover of marine biotas, as occurred at previous events that caused major perturbations of ocean chemistry [15–17]?

This question reveals a major paradox in the geologic record. Through the Phanerozoic, extended Greenhouse World times of higher CO_2 (Silurian–Devonian and Jurassic–Cretaceous) were characterized by prolific accumulations of biogenic carbonates, especially those associated with cyanobacterial and algal calcification (Figure 2).

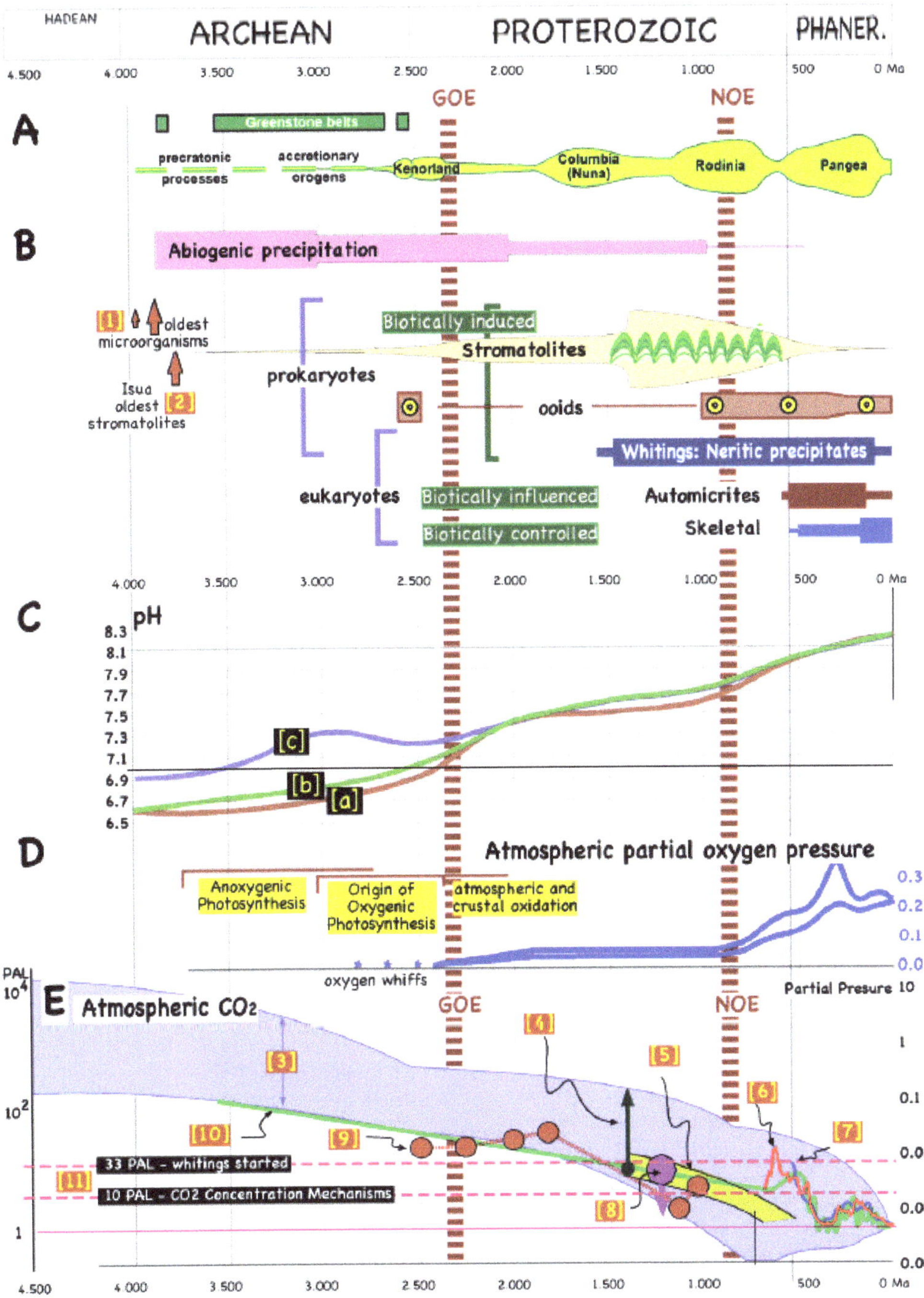

Figure 2. Carbonate precipitation nodes, genetic macroevolution rates, and estimates of oxygen and CO_2 in the atmosphere. (**A**): Supercontinents and crustal growth; precratonic, accretionary orogens and supercontinents assembly (redrawn from multiple sources). (**B**): Carbonate precipitation modes

through Earth's history (multiple sources). [1]: Oldest fossils [18]. [2]: Isua oldest stromatolites [19]. (**C**): pH evolution from [20]; [**a**]: nominal model, in which the median Archean surface temperature is slightly higher than modern surface temperatures. Archean land fraction was anywhere between 10% and 75% of modern land fraction; [**b**]: no Archean land endmember scenario; [**c**]: model with assumed 100 ppm Proterozoic methane and 1% Archean methane levels. (**D**): Maximum and minimum estimates of atmospheric partial oxygen pressure (after [21]). (**E**): Estimates of atmospheric CO_2 concentration; [3] from [22], upper and lower boundaries reflect average surface temperature for an ice-free (20 °C) and ice covered (5°) Earth; [4] Acritarch isotopic composition [23]; [5] C-isotope reservoir modeling [24]; [6] Phanerozoic GEOCARB III models [25]; [7] Royer compilation [26]; [8] from [27]; [9] paleosoil mass balances [28]; [10] from [29]; [11] Picoplanktonic whiting and partial sheath calcification commenced 1400–1300 Ma ago (33 CO_2 PAL), and cyanobacteria CCMs were induced when pCO_2 = 10 PAL [30]. PAL: present atmospheric level; GOE: the Great Oxygenation Event; NOE: Neoproterozoic oxygenation event. (This figure is an original compilation and interpretation of data by L. Pomar, based upon numerous sources).

For example, during the Cretaceous, micritic and algal (e.g., coccoliths) production was prolific. The emergence of rudists as major metazoan producers of skeletal carbonates was notable, and the "elevator" nature of many lineges has been postulated to have been the response to rapid inundation by carbonate muds. In the Cenozoic, the Paleogene was a time of transition. Coccolithophores and planktic foraminifers emerged as major producers of pelagic carbonate sediments through the Mesozoic. Following the Cretaceous-Paleogene extinction, new lineages of coccolithophores and planktic foraminifers diversified, while calcareous macroalgae and larger benthic foraminifers were notable producers of neritic carbonates. Coral reefs emerged as major carbonate factories as Icehouse World conditions developed in the Oligocene and became predominant in the Neogene, a conundrum noted more than 45 years ago [31].

Another paradox associated with changes in ocean chemistry over the course of Earth history relates to another essential element for life, which is commonly the most limiting nutrient required for essential processes including photosynthesis, growth and reproduction. Bioavailable phosphorus (P), typically occurring as phosphate (PO_4^{3-}), provides the backbone of nucleotides (DNA and RNA) and phospholipid membranes of cells, and is used in many other essential functions including energy storage and transfer associated with adenosine triphosphate (ATP), and for the synthesis of proteins and enzymes. Kempe and Degens [32] postulated that abundant dissolved PO_4^{3-} in the Archaean seas helped to foster the evolution of life (see also [33,34]).

2. The Hypothesis

The goal of our paper is to explore the hypothesis that decline in both CO_2 and PO_4^{3-}, and the history of changes in modes of calcification through Earth history, are tied to the close and complex relationships between biogenic carbonates and phosphate availability. The **Phosphate Extraction Mechanism (PEM)** provides an explanation for how calcifying microbial and algal biota can thrive in oligotrophic conditions. During daylight, calcification can be coupled to photosynthesis. With energy from active ion transport, protons can be split from bicarbonate, providing a carbonate ion (CO_3^{2-}) for calcification and a CO_2 molecule for photosynthesis. Phosphate is adsorbed during calcium carbonate ($CaCO_3$) precipitation. At night, respiration releases CO_2 and can promote partial dissolution of diurnally precipitated carbonates. Phosphate that is adsorbed onto precipitating $CaCO_3$ during daylight can be desorbed at night, making it available for uptake by the calcifying cyanobacteria or algae.

3. The Hadean–Archaean

Conditions on the Hadean Earth were clearly very different from present, and even from those of the Proterozoic. One hypothesis is that the early Earth was hot, following the moon-forming impact, and cooled to the point where liquid water was present after about 10 million years [35,36]. Whether the Earth's surface continued to be thermophilic well into the Archaean is still debated, though ~4.3 Ga rocks near Hudson Bay are suspected to have formed under warm, greenhouse conditions [37].

Over the past 4.5 Ga, heat-, gravity-, and tectonically driven processes have concentrated Iron (Fe) and Nickel (Ni) in the Earth's core. The dominant elements in the Earth's crust are Oxygen (O), Silicon (Si), Aluminum (Al), Iron (Fe), Calcium (Ca), Magnesium (Mg), Sodium (Na), and Potassium (K). In the mantle, Mg is nearly an order of magnitude more prevalent than in the crust, where proportions especially of Si and Al are higher. Basaltic, komatiitic (ultramafic) and mantle ejecta are potent CO_2 sinks (all are rich in Mg^{2+} and Fe^{2+}, with somewhat less Ca^{2+}) and their subduction may have drawn down atmospheric CO_2 [36,38]. Alkalic rocks in India indicate carbonate subduction occurred by 4.26 Ga [36].

Through the Archaean, the combination of high atmospheric CO_2 concentrations and the prevalence of Mg^{2+}, Fe^{2+}, and Ca^{2+} resulted in formation of massive abiogenic dolomites and limestones, indicating extreme oversaturation of waters [39] (Figure 2B). The earliest hints of life date back ~3.8 billion years (early–mid Archaean, Figure 2A,B), and by ~3.7 Ga, some metacarbonate rocks, found in the Isua supercrustal belt in southwest Greenland, contain 1–4-cm-high isolated and aggregated stromatolites [19]. Through the Proterozoic, massive abiogenic carbonates declined in prevalence relative to biogenic carbonates and are unknown in modern marine environments, even in those that are strongly oversaturated with respect to $CaCO_3$. In a water body containing abundant Ca^{2+} and dissolved inorganic carbon, modest increases in saturation, whether by warming, reduced hydrostatic pressure, evaporation, or by biogenic processes such as photosynthetic uptake of CO_2 or microbial sulfate reduction, can trigger $CaCO_3$ or $CaMg(CO_3)_2$ precipitation.

So the questions we pose are these: In post Archaean oceans, did biogenic processes become so ubiquitous that they became the drivers and controllers of $CaCO_3$ precipitation, resulting in the overwhelming predominance of biogenic carbonate production? Was the sequence of carbonate factories through geologic history, from predominantly abiogenic precipitation in the Archaean, to predominance of biologically induced geochemical precipitation in the Proterozoic, to the addition of biologically controlled calcification during the Phanerozoic (Figure 2B), consistent with and driven by evolutionary processes involving luminosity of the Sun, the stratification of the Earth, and the emergence of life, with their combined effects on ocean chemistry (Figure 2C)?

The luminosity of the Sun has increased ~30% during the past 4.5 Ga [40]. The composition of the Earth's atmosphere has changed (Figure 2D,E), with substantial decline in greenhouse gases, including CO_2 (Figure 2E) and methane (CH_4). For example, CO_2 has been estimated to have been ~100 times higher in the Hadean than modern levels and ~10 times higher in the Cambrian than present [36]. As noted above, ultramafic ejecta are potent CO_2 sinks and their subduction may have drawn down atmospheric CO_2 in the Hadean [38].

4. Pertinent Chemical Processes

Before delving into biological processes, we briefly examine some chemical factors associated with CO_2 in aqueous environments (Figure 3). Two measures associated with but not restricted to CO_2 are pH and alkalinity. Both are important in water chemistry, but their interactions can be confusing when considering calcification (29,38,41). The concentration of hydrogen ions (H^+), indicating how acidic or basic a substance is, is measured as pH. Alkalinity indicates the ability of a solution to neutralize acids (i.e., buffering capacity). Alkalinity consists of ions that can incorporate H^+ (protons) into their structure, limiting the availability of those protons that would otherwise lower pH. Carbonate alkalinity measures the concentrations of $CO_3{}^{2-}$ and $HCO_3{}^-$ ions, which are typically present in the highest concentrations in natural waters. Total Alkalinity reflects primarily $CO_3{}^{2-}$ and $HCO_3{}^-$ ions, but also includes $PO_4{}^{3-}$, borate, orthosilicate, sulfides, and organic acids.

Shallow water bodies can be strongly influenced by rising temperature and evaporation, concentrating the alkaline ions and promoting the precipitation of carbonate minerals, assuming that appropriate cations are available (e.g., Ca^{2+}, Mg^{2+}). Moreover, any process that takes up protons can also promote such precipitation.

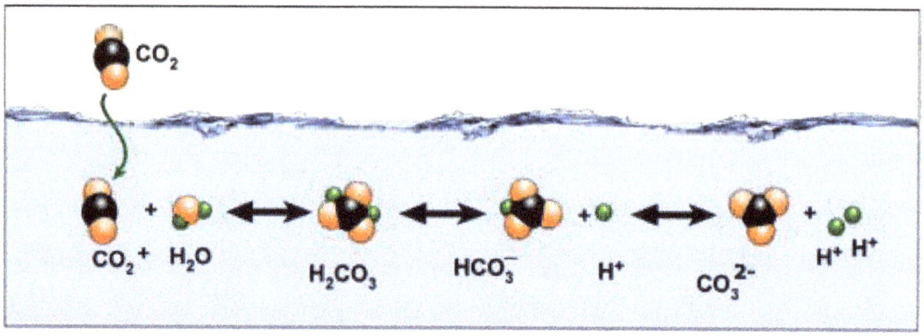

Figure 3. In an aqueous solution, carbonate, bicarbonate, carbon dioxide, and carbonic acid exist together in dynamic equilibrium, e.g., [10,20]. When CO_2 is absorbed by seawater, a series of chemical reactions increase the concentration of H^+ and cause the seawater acidity to increase. When calcium carbonate (e.g., limestone) reacts with acidic free hydrogen (H^+) ions in seawater, the solid calcium carbonate dissolves, releasing free calcium (Ca^{2+}) ions and free bicarbonate (HCO_3^-) ions. (This figure is an original interpretation by L. Pomar based on numerous sources).

As elements that are relatively abundant in both crustal and mantle rocks, Mg^{2+}, Fe^{2+}, and Ca^{2+} ions are soluble in aquatic environments. Thus, the combination of high carbonate alkalinity and abundant reactive cations, especially in warm waters, provided prolific sources of ions to produce abiogenic carbonates, providing critical storage of atmospheric CO_2 on geological time scales [41].

5. Pertinent Biological Processes

All living organisms carry traces of the histories of their ancestors within their genetic makeup and, in the case of the Eukarya, literally within their cells. The diversity of metabolic pathways in the prokaryotic Archaea and Eubacteria far surpasses the comparatively limited pathways found in the Eukarya, which are metabolically limited to those of their Proteoarchaeota and α-proteobacterial symbiotic predecessors [42].

Early microbial forms evolved in aquatic environments lacking free oxygen (O_2). Autotrophic processes in aquatic systems require: (a) an energy source, (b) source of dissolved inorganic carbon (DIC), (c) a source of protons, and (d) nutrients (fixed N, P, Fe, and trace elements) necessary for cell growth and reproduction [43]. Energy sources include oxidation of inorganic compounds (e.g., H_2S, CH_4) by chemoautotrophs and sunlight by photoautotrophs [44]. The primary sources of DIC are CO_2 or CH_4. In photoautotrophy, proton donors can be H_2, Fe^{2+} or H_2S, which do not release O_2 (anoxygenic photosynthesis), or H_2O that releases O_2 as a byproduct (oxygenic photosynthesis).

RuBisCO is the enzyme that catalyzes the first major step of carbon fixation in the Calvin cycle the process of photosynthesis. As the most abundant protein on Earth, RuBisCO is found in all three domains of life (Eubacteria, Archaea and Eukarya), and fixes more than 90% of the inorganic carbon that is converted into biomass [45]. However, this enzyme evolved in an O_2-free atmosphere, prior to the Great Oxygenation Event, and does not efficiently discriminate between CO_2 and O_2. Enzyme activity and specificity are reciprocally linked: faster RuBisCO has a higher error rate and more specific RuBisCO has a lower catalytic rate [45].

6. CCMs: CO_2 Concentrating Mechanisms

It may seem paradoxical that RuBisCO, this ubiquitous and essential enzyme, has not become more efficient or been replaced by more efficient enzymes. However, CCMs (CO_2 concentrating mechanisms) evolved instead by adaptation of active transport processes and compartmentalization of accumulated HCO_3^- [46]. The CCMs in cyanobacteria and microalgae enhance photosynthetic carbon fixation by energy-driven uptake of HCO_3^- and its conversion to CO_2. The minimum requirements for a cyanobacterial CCM are one energy-driven active transporter that accumulates HCO_3^- in the cytosol, and an anion-permeable carboxysome containing intracellular carbonic anhydrase [47]. Central to the functioning of the cyanobacterial CCM is the carboxisome (Figure 4), a protein micro-compartment within the cell that contains RuBisCo and a carbonic anhydrase. The latter converts HCO_3^- into CO_2 within the carboxysome, concentrating CO_2 up to 1000-fold around the active site of RuBisCO [46], thus overcoming the inefficiency of RuBisCO [24] and its inability to distinguish O_2 from CO_2.

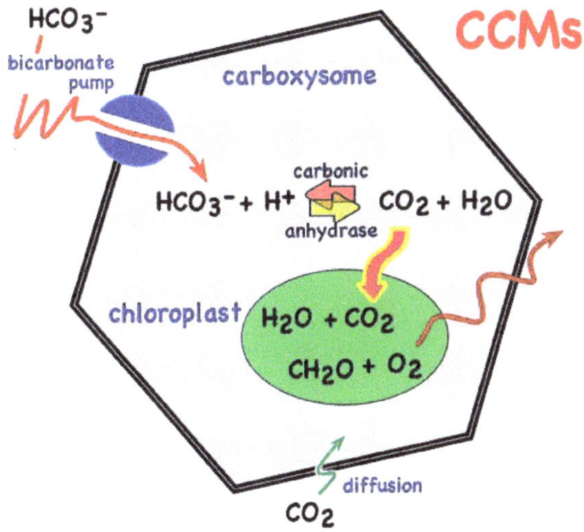

Figure 4. Central to the cyanobacterial CCM (CO_2 Concentrating Mechanism) is the carboxisome, a cellular micro-compartment containing RuBisCo and a carbonic anhydrase (CA) that, by converting HCO_3^- into CO_2, concentrates CO_2 as much as 1000-fold around the active site of RuBisCO [46]. (This figure is an original interpretation by L. Pomar based on several sources).

For eukaryotic algae with CCMs, the presence of pyrenoids is often associated with the occurrence of a CCM, although there are exceptions [47]. Eukaryotic algal CCMs likely originated independently in different clades of algae, as indicated by the diversity of inorganic C transporters and of carbonic anhydrases, as well as the lack of a strict correlation between occurrence of a CCM and the presence of a pyrenoid [47].

7. Phosphorous

Another essential element for life forms that has fluctuated downward in bioavailability through Earth history is phosphorous (P). Ranking 11th among the elements in the Earth's crust, P occurs primarily as phosphate (PO_4^{3-}) in apatite. Although found in most igneous and metamorphic rocks, mafic rocks commonly contain at least an order of magnitude more apatite than most felsic rocks (granites and rhyolites) [29]. Thus, as Fe- and Mg rich materials have been concentrated in the Earth's mantle and core, so too has P. By the late Archean, continental-scale granitic cratons had developed on the Earth's crust, contributing to the decline in PO_4^{3-} in aquatic environments [32].

Phosphorus is one of the eleven macro-biogenic elements that make up living organisms [C, O, H, N, S, P, Na, K, Ca, Mg, Cl, Fe]. Bioavailable PO_4^{3-} is essential for all forms of life, although it has received less attention than C, N, and Fe in the evolution of autotrophy in relation to environmental changes during the last 4 Ga [48–50]. As noted in the Introduction, PO_4^{3-} is essential to cellular structure and processes, including nucleotides (DNA and RNA), phospholipid membranes, energy transfer and storage, and for synthesis of proteins and enzymes [32]. However, PO_4^{3-} became one of the most limiting nutrients for primary productivity in Phanerozoic marine ecosystems due to its low solubility in oxygenated waters and because, unlike bioavailable dissolved N (DIN), PO_4^{3-} cannot be fixed from the atmosphere [34,48,51].

Microbial uptake of PO_4^{3-} takes place through at least two kinetically distinct processes. A "low affinity component" operates continuously, apparently driven by proton-motive force. A "high affinity component" requires energy from ATP and operates when internal PO_4^{3-} pools are depleted. Was the decline in available PO_4^{3-} in Proterozoic oceans related to changes in major precipitation modes of $CaCO_3$?

A key factor in the bioavailability of PO_4^{3-} in the Archean and early Proterozoic oceans was the lack of free oxygen [42,52]. By the early Proterozoic, cyanobacteria were producing free oxygen, triggering the Great Oxidation Event (~2.4–2.1 Ga) (Figure 2). With the emergence of O_2 into the atmosphere, terrestrial weathering dramatically changed, resulting in oxidation of sulfides ubiquitous in continental rocks. A major result was the delivery of substantial amounts of sulfates to the oceans, and those dissolved sulfates provided the oxidation potential for sulphate-reducing microbes, which are obligate anaerobes. The result was an oxygenated mixed layer/photic zone to perhaps 200 m depth, but the vast ocean depths remained anoxic and sulphidic (i.e., euxinic) through most of the Proterozoic [42,52].

The bioavailability of PO_4^{3-} declined in surface waters, based on the relative insolubility of PO_4^{3-} in oxygenated waters compared to its much greater solubility in euxinic waters [51,52]. Shallow waters exposed to sunlight became depleted in PO_4^{3-} as a consequence of reduced solubility in terrestrial-runoff waters and in oxygenated surface waters, and were further diminished by biological uptake by photosynthetic cyanobacteria and microalgae, and by geochemical uptake into ferric oxyhydroxides. At the same time, the much greater volumes of subsurface, aphotic waters, which were euxinic and therefore in which PO_4^{3-} was highly soluble, became major repositories. In that respect, like today, hydrodynamic processes such as mixing and upwelling must have been critical mechanisms for supplying PO_4^{3-} to photosynthesizing microorganisms in the surface waters [51].

By the late Neoproterozoic, sufficient photosynthetic oxygen production had occurred to oxidize most oceanic waters. Two additional contributors to the increase in atmospheric O_2 and dissolved O_2 throughout the oceans were primary production by eukaryotic algae and increased solubility of O_2 in cold oceanic waters associated with Cryogenian glaciation. The latter also would have increased the rates of ocean turnover and delivery to surface waters of dissolved PO_4^{3-} from subsurface waters (Figure 5). Shen et al. [53] postulated that the oxygenation of the oceans allowed diverse animal lineages to increase in size.

On scales of major plate-tectonics processes, massive basalt flows and tectonic uplift increase delivery of weathering-mobilized PO_4^{3-} to the oceans [48,49]. On intermediate time scales, changes in rates of deep-ocean circulation influence rates of delivery of DIN and PO_4^{3-} to surface waters, where they promote primary productivity [51,54]. In some cases, regional changes in ocean circulation have brought PO_4^{3-} rich subsurface waters into oxygenated conditions, creating major phosphatization events that have resulted in economically valuable phosphate ore deposits such as the prolific Lutetian-Messinian phosphorites of Morocco [55]. Most PO_4^{3-} in the ocean is cycling within the biotic system, and is returned to seawater by microbial activity. Only a small fraction of the phosphorus is buried in sediments, as reduced conditions within the sediments promotes dissolution of PO_4^{3-} into sediment pore waters and its return to ocean cycling [54].

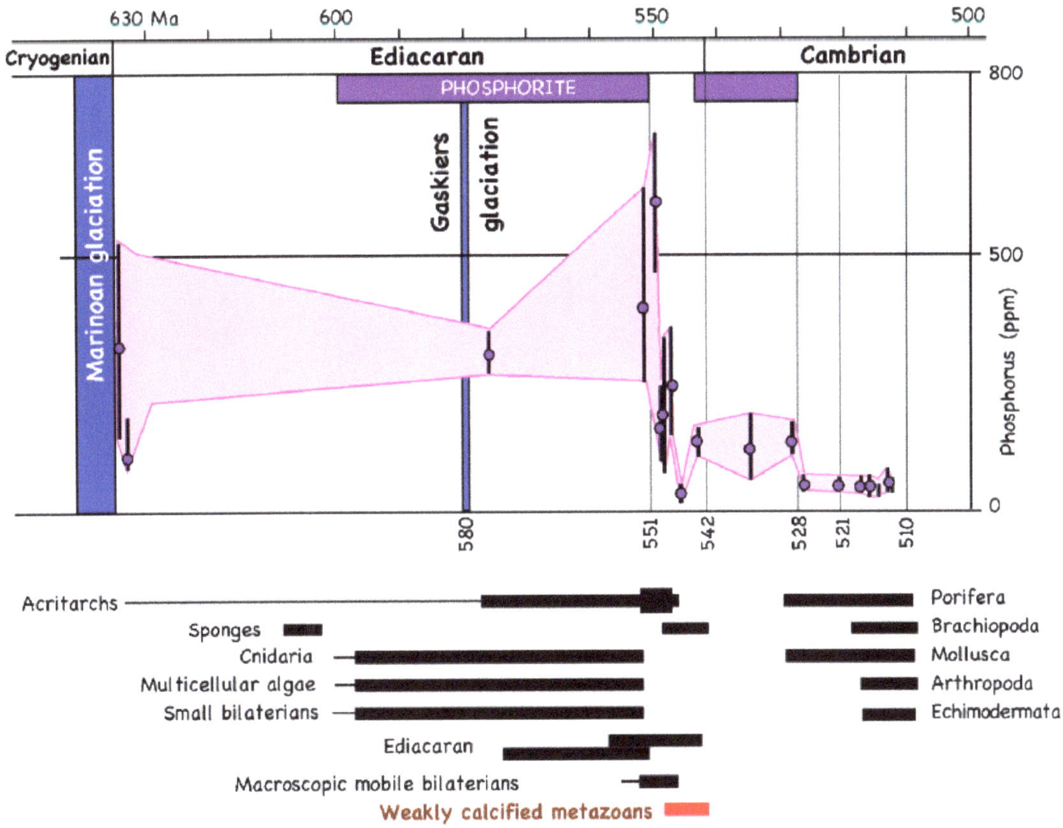

Figure 5. Evolution of Phosphorous (P) concentration through the late Neoproterozoic and Cambrian and its relationship with the emergence of major groups of organisms. Notice that the first weakly calcified metazoans (red bar) appeared during a minimum of P at the end of the Ediacaran. Modern lineages of metazoans with hard skeletons appeared by mid-Cambrian during another minimum in P concentration. (This figure is an original interpretation by L. Pomar based on numerous sources).

8. Calcification as a Sink and a Source of Both CO_2 and PO_4^{3-}

Precipitation of carbonate minerals is a long-term sink for CO_2, but in the short term, can serve as a source (e.g., [41] and references therein). Calcification typically involves two bicarbonate ions, incorporating one CO_3^{2-} into the $CaCO_3$ mineral structure and releasing the other into the environment as CO_2 [56]. In aquatic environments, photosynthetic uptake of CO_2 or HCO_3^- reduces availability of CO_2, increases pH and can lead to oversaturation with respect to calcite. Dissolved PO_4^{3-} is adsorbed onto and co-precipitates with calcite, with incorporation of some of the surface PO_4^{3-} into the bulk structure as crystal growth proceeds (Figure 6). However, higher concentrations of dissolved phosphate inhibit the growth of calcite crystals and can stop growth completely [57].

Numerous studies have documented removal of PO_4^{3-} associated with $CaCO_3$ precipitation, for example during dense blooms of some cyanobacteria. Guldbransen and Cremer [58] experimentally demonstrated co-precipitation of $CaCO_3$ and PO_4^{3-}, and Kitano et al. [59] found that stirring increases the amount of PO_4^{3-} that co-precipitates. Millero et al. [60] demonstrated that PO_4^{3-} uptake on $CaCO_3$ minerals is a multistep process. Depending upon water chemistry and temperature, adsorption or desorption can occur rapidly, in minutes to a few hours, with much slower processes lasting more than one week. Up to 80% of the adsorbed PO_4^{3-} can be released from $CaCO_3$ over one day. The amount of PO_4^{3-} left on the $CaCO_3$ is close to equilibrium adsorption [60]. Thus, the adsorption of PO_4^{3-} during $CaCO_3$ precipitation can be both a sink and a source of phosphates (Figure 7).

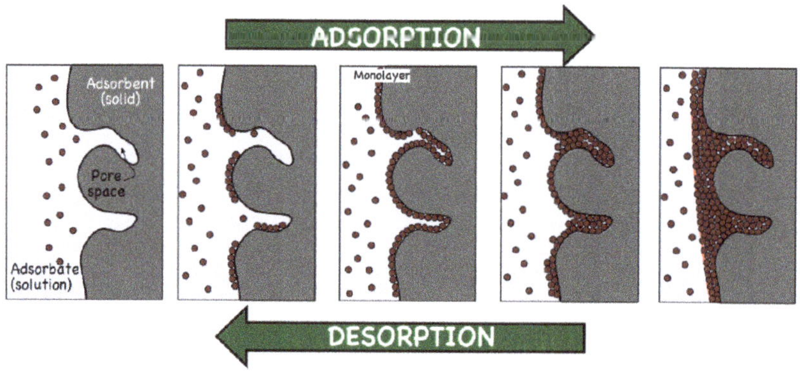

Figure 6. Simplified diagram of the adsorption process (adapted from [58–60]). Atoms, molecules or ions adhere to a surface by weak residual forces (e.g., van der Waals, electrostatic). It can be multilayer or monolayer. (This figure is an original interpretation by L. Pomar based on several sources).

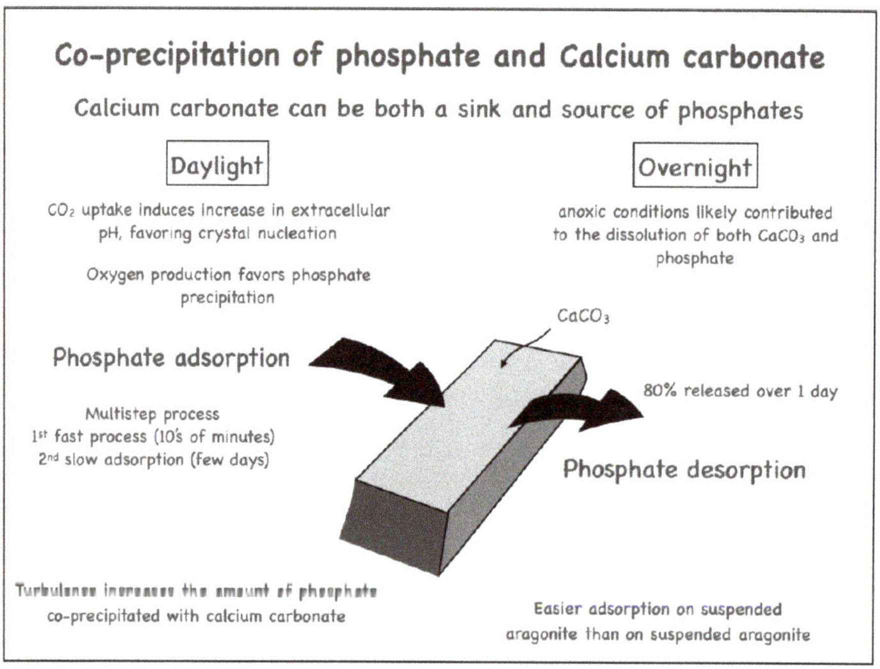

Figure 7. Co-precipitation of phosphate and calcium carbonate; note that phosphate is more readily adsorbed onto and released from aragonite than onto/from calcite ((This figure is an original interpretation by L. Pomar based on several sources, including [58–60]).

9. Prokaryotic Organo-Sedimentary Systems

Stromatolites, the fossil evidence of calcified microbial mats, can be viewed as the first biotic process to produce a hard body [61]. By the Late Archaean, ooids—the second biotic process to produce a hard body—appeared when photosynthetic O_2 production by cyanobacteria was occurring. What might have been the advantage for these benthic microbial systems to produce coated-mineral structures? Such coatings would have restricted light penetration, as well as diffusion of DIC and nutrients required for photosynthesis, growth and reproduction. Furthermore, what could be the advantage of inducing carbonate-mud precipitates (whitings) by neritic photosynthetic prokaryotes (picoplankton) and eukaryotes (phytoplankton)? This third microbially mediated organo-sedimentary system to produce carbonate precipitates emerged by the Mesoproterozoic, following the Great Oxygenation Event (GOE) (Figure 2).

Cyanobacteria have remarkably few nutritional requirements. They use light as source of energy, H_2O or H_2S as electron donors and CO_2 (or HCO_3^-) as the source of inorganic carbon [62]. Additionally, many species can fix atmospheric N_2 to NH_4^+, which makes them independent of DIN sources such as nitrate, ammonium or organic nitrogen. Hence, the most critical nutrients for cyanobacteria are PO_4^{3-}, which can be stored intracellularly as polyphosphate, and, to a lesser extent, Fe^{2+}. Both are limited by their reactivity with O_2. Moreover, cyanobacteria can minimize their PO_4^{3-} requirements through the synthesis of sulfolipids [63], allowing them to colonize low-nutrient environments, as demonstrated by their ubiquity in both terrestrial and aquatic environments.

Thus, precipitation of $CaCO_3$ during the day augments CO_2 availability for photosynthesis [64], while PO_4^{3-} adsorption sequesters that scarce but essential nutrient. At night, desorption of PO_4^{3-} makes it available for uptake by the cells. In microbial mats (Figure 8) in darkness, anoxic conditions resulting from continued sulfate reduction, aerobic respiration and sulfide oxidation likely contribute to the dissolution of $CaCO_3$ and increase the potential for PO_4^{3-} uptake by the microbial assemblage.

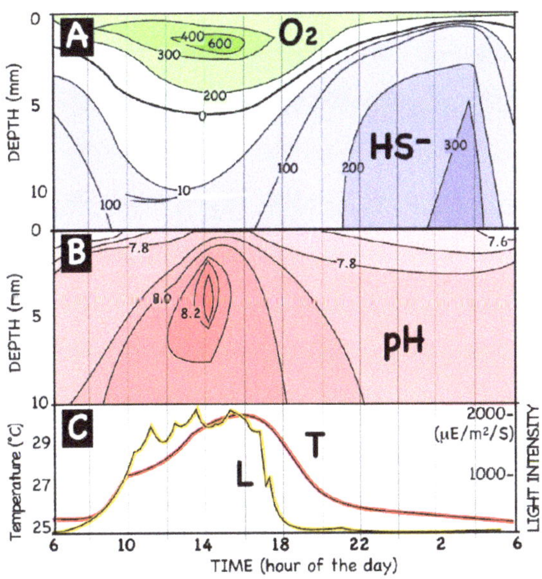

Figure 8. Oxygen and sulfide concentration (ppm) in (**A**) and pH in (**B**) variation during a diel cycle in the upper 12 mm of modern marine stromatolites in the Exuma Cays, Bahamas. (**C**): Light intensity and temperature at the surface of the stromatolite. (Adapted with permission from Ref. [65]. Copyright ©2002, American Chemical Society.).

Giant ooids, associated with cyanobacterial mats and stromatolites, are common in Late Archaean rocks. They have been described in the 2.64 Ga old Ghaap Group in South Africa [66] and in the 2.72 Ga old Pilbara group in Australia [67], becoming abundant and characteristic of many Neoproterozoic carbonates [68]. Microbially mediated organo-mineralization processes are likely involved in the origin of ooids. They can either be biologically induced (e.g., by-products of metabolic activities that increase environmental alkalinity and trigger carbonate precipitation), or biologically influenced (e.g., microbial extracellular polymeric substances can serve as templates for carbonate mineralization [69]). Again, the advantage for the microbial consortium to precipitate $CaCO_3$ coatings around particles was likely in PO_4^{3-} adsorption, associated with the precipitation of metastable amorphous carbonate [70,71] (Figure 9).

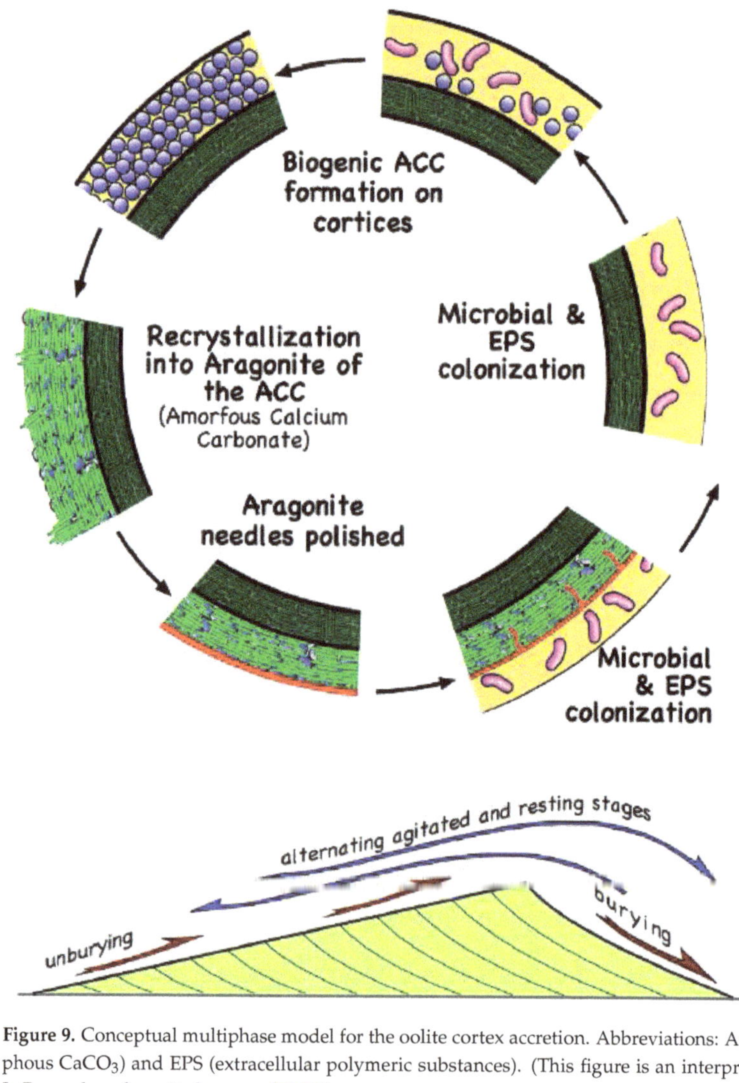

Figure 9. Conceptual multiphase model for the oolite cortex accretion. Abbreviations: ACC (amorphous $CaCO_3$) and EPS (extracellular polymeric substances). (This figure is an interpretation by L. Pomar based on cited sources [70,71]).

A significant source of lime-mud production in modern oligotrophic shallow marine and lacustrine environments also is linked to microbial calcification [72]. Whitings are the drifting milky clouds that are commonly observed in warm, shallow seas. Robbins et al. [73] estimated that whitings production might account for more than 40% of the bank-top- and peri-platform Holocene muds on the west side of Great Bahama Bank. The production of whitings is linked to blooms of cyanobacteria and planktic green algae in oligotrophic aquatic environments via CO_2 uptake [74]. Photosynthetic carbon fixation during blooms induces an increase in extracellular pH, favoring crystal nucleation [72]. What would it be the advantage for photosynthetic pico- and microplankton to induce carbonate-mud precipitation in the water? Blooming means that CO_2 and nutrient acquisition must keep pace with the increase of organic matter required for cell division. With the diurnal cycle of rapid adsorption and desorption of PO_4^{3-}, the milky cloud can efficiently capture this scarce nutrient, which would be of particular advantage in extremely oligotrophic conditions.

10. Biocalcification in Photosynthetic Eukaryotes

The basic eukaryotic cell is descended from an anaerobic Proteoarchaeota that engulfed purple, non-sulfur α-proteobacteria that could utilize oxygen but were not obligate aerobes [42,44]. All algae have that basic eukaryotic cell and all algal plastids are believed to have originated from cyanobacterial endosymbionts [42,75]. Molecular phylogenetic studies have demonstrated that the morphological diversity of the algae results from their polyphyletic origins within the Eukarya [75,76]. An interesting question then, is: Why do some marine algae within different phylogenetic groups invest energy into producing $CaCO_3$ skeletal structures, whereas most do not?

Tropical macroalgae cope with three strong selective pressures: limited access to CO_2/HCO_3^- due to warm water and competition among photosynthetic organisms, limited access to PO_4^{3-}, and herbivory. Calcification is clearly an anti-herbivory defense. Furthermore, there are striking differences in percent tissue PO_4^{3-} between calcified and fleshy algae, indicating different nutrient-acquisition strategies. The effects of nutrient enrichment with nitrogen (DIN) and PO_4^{3-} on productivity and calcification of fleshy and calcareous algae differ. DIN and PO_4^{3-} frequently enhance productivity of fleshy algae, but do not increase calcification rates of calcareous species [77].

Halimeda, a codiacean macroalga common in warm, shallow-marine environments, can add up to one segment per day, per branch (Figure 10). However, the algae require nutrient uptake for growth and reproduction in warm waters where PO_4^{3-} levels are often below detection limits. Calcification occurs during daytime within an organic matrix in the outer utricle walls of one-day-old segments. Photosynthetic removal of CO_2 by diffusion across a membrane and enzymatic anhydrase activity induces aragonite precipitation in the utricle space [78,79]. At night, respiration elevates CO_2 and decreases CO_3^{2-} saturation, which decreases seawater pH in the utricle space. The aragonite needles partially dissolve, break and recrystallize into micro-anhedral crystals (<1 μm). Such partial dissolution at night would facilitate desorption of PO_4^{3-} from the skeleton.

Coralline red algae occur as thin crusts on hard substrata that can produce free-living rhodoliths, or as thalli with articulated branches. Most precipitate high-Mg calcite rather than aragonite. Calcification dynamics (Figure 11) depend on nutrient concentrations; PO_4^{3-} inhibits the crystalline lattice and hampers calcite precipitation [80]. Enrichment of PO_4^{3-}, along with other nutrients, favors the growth of macroalgae and phytoplankton, which reduces water transparency and impedes coralline algal growth. Calcification associated with photosynthesis exhibits day-night cyclicity that is also recorded in diel pH fluctuations. Abiotic re-precipitation of $CaCO_3$ can occur in crustose-coralline red algae during nighttime [81] and pH drop at the boundary layer in darkness has been documented [82]. Thus, partial dissolution occurring at night would provide favorable conditions to remobilize adsorbed PO_4^{3-} from the calcitic skeleton and facilitate uptake by the algal cells.

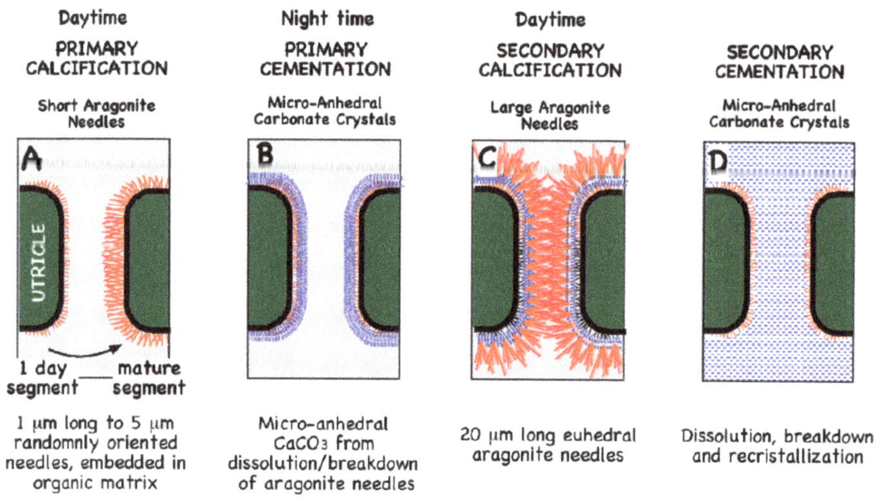

Figure 10. Calcification stages of *Halimeda* [78,79]. (**A**): Growth of short skeletal needles on the utricle wall; (**B**): recrystallization of short needles to micron-sized anhedral crystals. (**C**): secondary calcification by long, dense, euhedral, skeletal-aragonite needles, (**D**): the primary inter-utricular space in the rim of the segment is filled with micron-sized anhedral crystals. (This figure is an original interpretation by L. Pomar based on cited sources [78,79]).

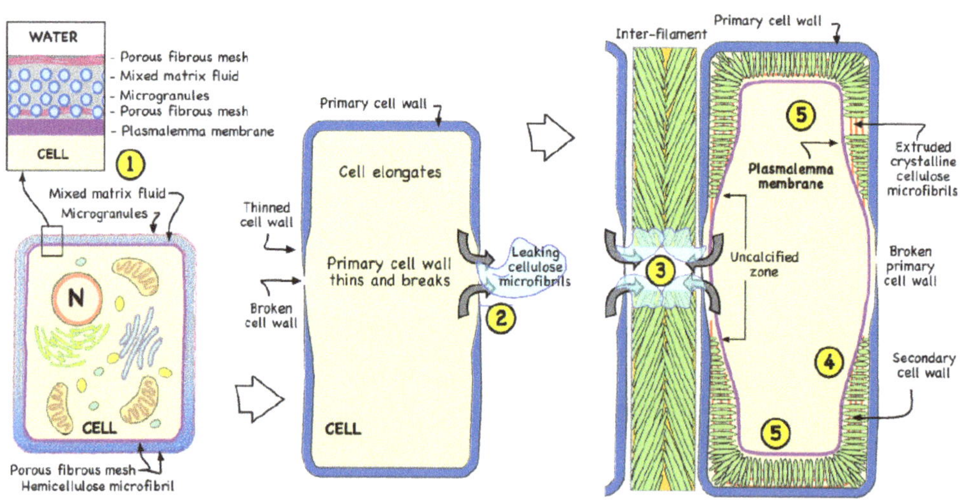

Figure 11. Skeletal formation in red algae. Process steps: (1) Primary cell wall is forming. Hemicellulose microfibrils act as nucleating substrate for micro-granules. (2) Cellulose microfibrils extrude from plasma membrane and leaks through broken primary cell wall. (3) Interfilament grains initiate attached to the external part of the primary cell wall and grow in the leaked cellulose microfibril. (4) Secondary cell wall thickens, extruded cellulose microfibrils stop leaking and mineralize as Mg-calcite. (5) Plasmalemma is pushed inward as cellulose microfibrils continue to extrude and mineralize. Secondary cell wall fully formed. (This figure is an original interpretation by L. Pomar based on sources including [80]).

Coccolithophores are photosynthetic protists that produce small calcitic disks (coccoliths). These oval-shaped plates of $CaCO_3$ consist of double discs, composed of radial arrays of minute, elaborately-shaped crystal units. They form delicate crystalline lace or open web-like patterns in which the rate of the $CaCO_3$ crystal surface/volume is very high. Originating by the Late Triassic (Figure 2), they proliferated during the Cretaceous, and, together with planktic foraminifera, allowed widespread carbonate production in the open ocean [83]. Coccolithophores can produce as much as two coccoliths per hour [84]. The variability of coccolith shapes indicates they serve diverse functions. Young [85] noted features that defy simple physical functions, particularly complex mesh structures and the variety of elaborate forms, and suggested that coccoliths might be adaptations for nutrient uptake.

Coccolithophores thrive in waters with minimal PO_4^{3-}, that is, insufficient to promote growth of other phytoplankton. Coccolith formation is less adversely affected by nutrient deficiency than is cell division and growth [86]. The nutrient requirements for the organic cellular components of a coccolithophorid cell are similar to those for non-calcifying phytoplankton. However, forming coccoliths requires minimal nutrient cost, as coccolith production continues even when cell division ceases because of nutrient limitation [87,88].

In modern *E. huxleyi*, coccoliths nucleate and grow in a Golgi-derived coccolith vesicle, from where they migrate to the cell surface (Figure 12). Within the coccolith vesicle, polysaccharides are thought to regulate nucleation and subsequent growth of coccoliths [89]. Sviben et al. [90] identified a reservoir compartment, distinct from the coccolith vesicle, with high concentrations of Ca^{2+} and PO_4^{3-}. Only Ca^{2+} is delivered to the coccolith vesicle [91], whereas PO_4^{3-} is diverted to the cell for growth and reproduction.

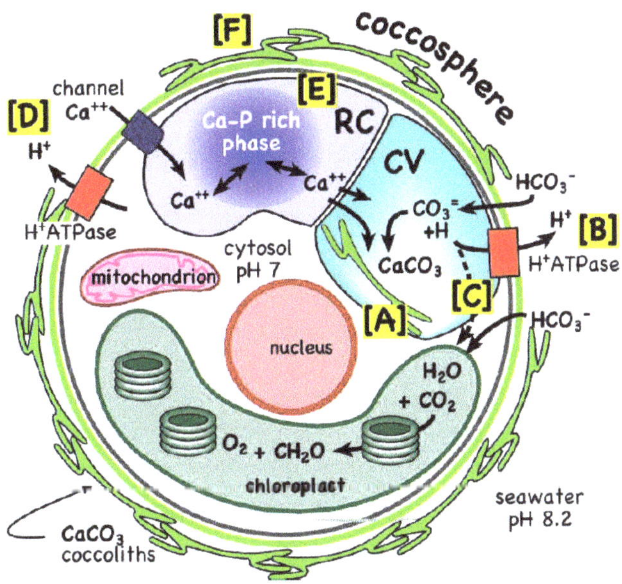

Figure 12. Calcification processes in coccolithophorids. **[A]** Coccolith formation occurs within the Golgi-derived coccolith vacuole (CV). **[B]** Intracellular $CaCO_3$ precipitation from bicarbonate (HCO_3^-) releases equimolar H^+ that must be rapidly removed from the CV to maintain pH for $CaCO_3$ precipitation. **[C]** Some H^+ may be utilized by photosynthesis. **[D]** Though not fully understood, it is hypothesized that Ca^{2+} is recruited through Ca-channels and V-type ATPase pumps. **[E]** A reservoir compartment (RC) concentrates Ca^{++} and PO_4^{3-}. **[F]** The coccolith-associated polysaccharide may drive the Ca^{2+} from the coccosphere into the RC, from where only Ca^{2+} is transported to the coccolith vesicle (CV) and PO_4^{3-} is diverted to the cell for algal growth and reproduction. Transfer of Ca^{2+} from the RC into the CV uses Ca^{2+} channels and transporters. (This figure is an original interpretation by L. Pomar based on [88–91] and other sources).

11. CO_2 and Carbonate Depositional History

In aqueous solutions, CO_3^{2-}, HCO_3^-, CO_2, and H_2CO_3 co-exist in dynamic equilibria (Figure 3). When CO_2 reacts with water (H_2O), it complexes to H_2CO_3, which dissociates to H^+ and HCO_3^-, thereby increasing the concentration of H^+ and cause the seawater acidity to increase (i.e., the pH to decline). When $CaCO_3$ (e.g., limestone) reacts with free hydrogen (H^+) ions in seawater, the solid $CaCO_3$ dissolves, forming free calcium (Ca^{++}) ions and free bicarbonate (HCO_3^-) ions. As noted previously, the concentrations of ions of carbonate, bicarbonate, phosphate, borate, orthosilicate, sulfides, and organic acids constitute alkalinity in aquatic environments, especially HCO_3^-. The consumption of protons results in decreased H^+ activity. Thus, $CaCO_3$ acts to neutralize or buffer the solution by consuming H^+.

In the modern world, burning of fossil fuels, combined with widespread changes in land use, have resulted in rapidly increasing concentrations of atmospheric CO_2 that are causing the decline in the pH of surface seawater (e.g., [6–11]). Will the progressive warming and acidification cause mass extinctions and evolutionary turnover of marine biotas, as has been documented by paleoclimatic evidence [15–17]?

From this question emerges another major "carbonate paradox". Contrary to predictions of increased carbonate dissolution in response to increasing CO_2, extended high CO_2 Greenhouse World periods during the Phanerozoic (Silurian–Devonian and Jurassic–Cretaceous) were characterized by thick and extensive accumulations of fine-grained carbonates associated with cyanobacterial and algal calcification, commonly baffled by skeletal carbonates (e.g., stromatoporids, corals, rudists) (Figure 2) ([92] and references therein). In the Cenozoic, coral reefs emerged as major carbonate factories as Icehouse World conditions developed and became predominant in the Oligocene through the Holocene, a conundrum noted by Frost many years ago [31].

The solution to this paradox lies in two parts. First, there is the "rate" factor. Mass extincton events in the Phanerozoic record have long be recognized by carbonate depositional hiatuses that indicate times of global carbonate dissolution [93]. Very rapid increases in atmospheric CO_2 (Figure 3) or other atmospheric compounds that unite with water to form acids (e.g., $2SO_2 + H_2O \rightarrow H_2SO_4 + 2H^+$), whether caused by massive extrusion of flood basalts, bolide impact, methane-hydrate release, extensive burning of fossil fuels, or some combination of events, can indeed trigger carbonate dissolution resulting in hiatuses of tens of thousands of years (Paleocene–Eocene) or hundreds of thousand years (Cretaceous–Paleocene) [94] and references therein) or millions of years (Permian–Triassic) [17].

However, eventually terrestrial weathering and dissolution under high atmospheric CO_2 over time increased alkalinity and restored the carbonate factory. The massive accumulations of Silurian–Devonian and Jurassic–Cretaceous carbonates occurred during times of relatively rapid sea-floor spreading, which elevated both CO_2 and Ca^{2+} concentrations, triggered warming temperatures, sea-level rise, and expansive areas of relatively shallow basins and shelves. Reduced land areas limited freshwater input of dissolved nutrients and warm waters reduced rates of deep-ocean circulation. During Greenhouse-world conditions, the trophic resource continuum expanded (compared to Icehouse conditions), with limited extremely rich regions of upwelling and expansive regions of extreme oligotrophy, the latter fostering extensive production of carbonates [95]. Elevated evaporation rates, relatively shallow seas, and high alkalinity favored calcifying cyanobacterial and algal taxa that could sequester scarce PO_4^{3-} while producing excess organic matter upon which calcifying animals (e.g., rudists) could also thrive [92]. The name Cretaceous was actually derived from the Latin *creta*, meaning chalk [96].

During daylight hours, in warm, shallow waters, abundant microalgae take up available DIN and PO_4^{3-}, while high rates of photosynthesis deplete immediate access to CO_2/HCO_3^-. Calcification associated with photosynthesis alleviates both deficiencies. Two HCO_3^- provides one CO_2 for photosynthesis and a CO_3^{2-} for calcification, somewhat alleviating local CO_2 depletion. At the same time, PO_4^{3-} adsorption occurs during daytime calcification and is partially desorbed at night, becoming available for uptake by the cyanobacteria or algae. In addition, some dissolution of the $CaCO_3$ releases HCO_3^-, and along with community respiration, results in early morning maximum concentrations of HCO_3^-. With the onset of light, photosynthesis makes energy available to CCMs, providing the primary producers with access to and storage of CO_2, which becomes increasingly unavailable over the course of the day.

Finally, why did extensive coral reefs only become prevalent with the onset of the Icehouse World climates [31]? Again, the answer is likely climatic and geochemical influences on biological processes [92,97–100]. High-latitude cooling, compression of tropical habitats, and increasing temperature gradients between high and low latitudes and between surface and deeper waters [97] were major factors. Coral communities flourished circumtropically beginning in the late Oligocene [31]. By the early Miocene, reefs and associated biota expanded latitudinal distributions by more than 10° north and south [98]. The expansion in reef-building capacity corresponded to increasing Mg/Ca ratios in seawater [99] and falling atmospheric CO_2 concentrations [100], both of which promoted aragonite precipitation in warm, tropical waters. The majority of extant Symbiodiniacea lineages diversified since Middle Miocene [101]. Global cooling, including much cooler subsurface waters, likely benefited the coral-Symbiodiniacea symbioses, which are sensitive to photo-oxidative stress under elevated temperatures [102]. The substantial increase in rates of circulation of deep, cold waters resulted in more "intermediate" conditions of oceanic oligotrophy [95]. Pomar and Hallock [92] further postulated that Neogene cooling supported the co-evolution and expansion of zooxanthellate corals and coralline algae into shallow, high-energy waters, where their carbonate production potential was highest. In such environments, both calcification and PO_4^{3-} extraction are also potentially optimized.

12. Conclusions

Photosynthesis typically involves daytime CO_2 uptake and O_2 release predominating over respiration; at night, respiration consumes O_2 and releases CO_2. The result is strong diel variation in both pH and dissolved O_2 concentrations in aquatic environments. In warm-water environments in which alkalinity is relatively high and PO_4^{3-} concentrations are minimal, these variations have the potential to promote precipitation of $CaCO_3$ and associated adsorption of PO_4^{3-} in daylight when photosynthesis is active. At night, lower pH and oxygen availability can promote some $CaCO_3$ dissolution and PO_4^{3-} desorption. The combination of calcification and PO_4^{3-} extraction occurs in photosynthesizing cyanobacteria, as well as in a diverse array of calcifying nanophytoplankton and calcareous macroalgae, indicating that this process allows both prokaryotic and eukaryotic photosynthetic organisms to thrive in warm, alkaline, oligotrophic waters. Concentrations of both atmospheric CO_2 and surface-ocean PO_4^{3-}, which are essential for biological productivity, have declined by at least 1–2 orders of magnitude over Earth history. **The Phosphate Extraction Mechanism**, associated with photosynthetically induced calcification, has played a major role in the production and accumulation of carbonates throughout much of the Proterozoic and the Phanerozoic. Recognition of this relationship helps resolve the apparent paradox that periods in the Phanerozoic when atmospheric CO_2 levels were considerably higher than present (as well as higher than predicted Anthropocene concentrations) were times of massive accumulation of carbonates, predominantly produced by photosynthesizing cyanobacteria and calcifying algae. This synthesis also stresses the importance of rates of change, as times of rapid increase in atmospheric CO_2 concentrations were associated with mass-extinction events characterized by global carbonate-depositional hiatuses, requiring hundreds of thousands or even millions of years for surface waters of the oceans to regain sufficient alkalinity to sustain accumulation of massive carbonates.

Author Contributions: Conceptualization: L.P., P.H., G.M.-V. and J.I.B.; Methodology: L.P., P.H., G.M.-V. and J.I.B.; Investigation: L.P., P.H., G.M.-V. and J.I.B.; Writing: Original Draft Preparation, L.P. and P.H.; Writing: Review and Editing, P.H. and G.M.-V. All authors have read and agreed to the published version of the manuscript.

Funding: JIB acknowledges funding from the Basque Government to the Research Group IT1602-22. L.P., P.H. and G.M.-V. participation did not involved external funding aside from their academic institutions.

Institutional Review Board Statement: Not applicable.

Informed Consent Statement: Not applicable.

Data Availability Statement: Not applicable.

Acknowledgments: Gabriel Moyà (UIB) for the seminal discussions on this topic, and to the volume editor, H. Westphal, for her patience and suggestions.

Conflicts of Interest: The authors declare no conflict of interest.

References

1. Selly, R.C.; Sonnenberg, S.A. *Elements of Petroleum Geology*, 3rd ed.; Academic Press: Boston, USA, 2015; pp. 1–507.
2. Kharecha, P.; Kasting, J.; Siefert, J. A coupled atmosphere–ecosystem model of the early Archean Earth. *Geobiology* **2005**, *3*, 53–76. [CrossRef]
3. Morse, J.W.; Arvidson, R.S.; Lüttge, A. Calcium carbonate formation and dissolution. *Chem. Rev.* **2007**, *107*, 342–381. [CrossRef] [PubMed]
4. Burton, M.R.; Sawyer, G.M.; Granieri, D. Deep carbon emissions from volcanoes. *Rev. Mineral. Geochem.* **2013**, *75*, 323–354. [CrossRef]
5. Global Carbon Budget 2021. *Earth Syst. Sci. Data* **2022**, *14*, 1917–2005. [CrossRef]
6. Keeling, C.D. The concentration and isotopic abundances of carbon dioxide in the atmosphere. *Tellus* **1960**, *12*, 200–203. [CrossRef]
7. Scripps CO2 Program. Global Stations CO2 Concentration Trends. 2022. Available online: https://scrippsco2.ucsd.edu/graphics_gallery/other_stations/global_stations_CO2_concentration_trends.html (accessed on 19 September 2022).
8. Harris, D.C. Charles David Keeling and the story of atmospheric CO_2 measurements. *Anal. Chem.* **2010**, *82*, 7865–7870. [CrossRef]
9. IPCC (Intergovernment Panel on Climate Change). Available online: https://www.ipcc.ch/about/ (accessed on 22 August 2022).
10. Kleypas, J.A.; Buddemeier, R.W.; Archer, D.; Gattuso, J.-P.; Langdon, C.; Opdyke, B.N. Geochemical consequences of increased atmospheric CO_2 on coral reefs. *Science* **1999**, *284*, 118–120. [CrossRef]
11. Caldeira, K.; Wickett, M.E. Anthropogenic carbon and ocean pH. *Nature* **2003**, *425*, 365. [CrossRef]
12. Wigley, T.M. A combined mitigation/geoengineering approach to climate stabilization. *Science* **2006**, *314*, 452–454. [CrossRef]
13. Wolf-Gladrow, D.; Riebesell, U.; Burkhardt, S.; Bijma, J. Direct effects of CO_2 concentration on growth and isotopic composition of marine plankton. *Tellus* **1999**, *51B*, 461–476. [CrossRef]
14. Orr, J.C.; Fabry, V.J.; Aumont, O.; Bopp, L.; Doney, S.C.; Feely, R.A.; Gnanadesikan, A.; Gruber, N.; Ishida, A.; Joos, F.; et al. Anthropogenic ocean acidification over the twenty-first century and its impact on calcifying organisms. *Nature* **2005**, *437*, 681–686. [CrossRef] [PubMed]
15. Veron, J.E.N. Mass extinctions and ocean acidification: Biological constraints on geological dilemmas. *Coral Reefs* **2008**, *27*, 459–472. [CrossRef]
16. Kiessling, W.; Simpson, C. On the potential for ocean acidification to be a general cause of ancient reef crises. *Glob. Chang. Biol.* **2011**, *17*, 56–67. [CrossRef]
17. Clarkson, M.O.; Kasemann, S.A.; Wood, R.A.; Lenton, T.M.; Dains, S.J.; Richoz, S.; Ohnemüller, F.; Meixner, A.; Poulton, S.W.; Tipper, E. Ocean acidification and the Permo-Triassic Mass Extinction. *Science* **2015**, *348*, 229–232. [CrossRef] [PubMed]
18. Dodd, M.; Papineau, D.; Grenne, T.; Slack, J.F.; Rittner, M.; Pirajno, F.; O'Neil, J.; Little CT, S. Evidence for early life in Earth's oldest hydrothermal vent precipitates. *Nature* **2017**, *543*, 60–64. [CrossRef]
19. Nutman, A.P.; Bennett, V.C.; Friend CR, L.; Van Kranendonk, M.J.; Chivas, A.R. Rapid emergence of life shown by discovery of 3700-million-year-old microbial structures. *Nature* **2016**, *537*, 535–538. [CrossRef]
20. Krissansen-Totton, J.; Arney, G.N.; Catling, D.C. Constraining the climate and ocean pH of the early Earth with a geological carbon cycle model. *Proc. Natl. Acad. Sci. USA* **2018**, *115*, 4105–4110. [CrossRef]
21. Holland, H.D. The oxygenation of the atmosphere and oceans. *Phil. Trans. R. Soc. B-Biol. Sci.* **2006**, *362*, 903–915. [CrossRef]
22. Kasting, J.F. Earth's early atmosphere. *Science* **1993**, *259*, 920–926. [CrossRef]
23. Kaufman, A.J.; Xiao, S. High CO_2 levels in the Proterozoic atmosphere estimated from analyses of individual microfossils. *Nature* **2003**, *425*, 279–282. [CrossRef]
24. Kah, L.C.; Bartley, J.K. Effect of marine carbon reservoir size on the duration of carbon isotope excursions. Interpreting the Mesoproterozoic carbon isotope record. *Geol. Soc. Am. Abstr. Programs* **2004**, *36*, 78.
25. Berner, R.A.; Kothavala, Z. GEOCARB III: A revised model of atmospheric CO_2 over Phanerozoic time. *Am. J. Sci.* **2001**, *301*, 182–204. [CrossRef]
26. Royer, D.L.; Berner, R.A.; Montañez, I.P.; Tabor, N.J.; Beerling, D.J. CO_2 as a primary driver of Phanerozoic climate. *GSA Today* **2014**, *14*, 4–10. [CrossRef]
27. Kah, L.C.; Riding, R. Mesoproterozoic carbon dioxide levels inferred from calcified cyanobacteria. *Geology* **2007**, *35*, 799–802. [CrossRef]

28. Sheldon, N.D. Precambrian paleosols and atmospheric CO_2 levels. *Precambrian Res.* **2006**, *147*, 148–155. [CrossRef]
29. Sleep, N.H.; Zahnle, K. Carbon dioxide cycling and implications for climate on ancient Earth. *J. Geophys. Res.* **2001**, *106*, 1373–1399. [CrossRef]
30. Riding, R. Cyanobacterial calcification, carbon dioxide concentrating mechanisms, and Proterozoic–Cambrian changes in atmospheric composition. *Geobiology* **2006**, *4*, 299–316. [CrossRef]
31. Frost, S.H. Cenozoic reef systems of Caribbean. Prospects for paleoecologic synthesis. *Am. Assoc. Petrol. Geol. Stud. Geol.* **1977**, *4*, 93–110.
32. Kempe, S.; Degens, E.T. An early soda ocean. *Chem. Geol.* **1985**, *53*, 95–108. [CrossRef]
33. Crockford, P.; Halevy, I. Questioning the paradigm of a phosphate-limited Archaean biosphere. *Geophys. Res. Let.* **2022**, *49*, e2022GL099818. [CrossRef]
34. Ingalls, M.; Grotzinger, J.P.; Present, T.; Rasmussen, B.; Fischer, W.W. Carbonate-associated phosphate (CAP) indicates elevated phosphate availability in Neoarchean shallow marine environments. *Geophys. Res. Let.* **2022**, *49*, e2022GL098100. [CrossRef]
35. Zahnle, K.; Arndt, N.; Cockell, C.; Halliday, A.; Nisbet, E.; Selsis, F.; Sleep, N.H. Emergence of a habitable planet. *Space Sci. Rev.* **2007**, *129*, 35–78. [CrossRef]
36. Sleep, N.H. The Hadean-Archaean environment. *Cold Spring Harb. Perspect. Biol.* **2010**, *2*, a002527. [CrossRef] [PubMed]
37. O'Neil, J.; Carlson, R.W.; Francis, D.; Stevenson, R.K. Neodymium-142 evidence for Hadean mafic crust. *Science* **2009**, *321*, 1828–1831. [CrossRef]
38. Sleep, N.H.; Zahnle, K.; Neuhoff, P.S. Initiation of clement surface conditions on the early Earth. *Proc. Natl. Acad. Sci. USA* **2001**, *98*, 3666–3672. [CrossRef]
39. Lowe, D.R. Abiological origin of described stromatolites older than 3.2 Ga. *Geology* **1994**, *22*, 387–390. [CrossRef]
40. Goldblatt, C.; Zahnle, K.J. Clouds and the faint young Sun paradox. *Clim. Past* **2011**, *7*, 203–220. [CrossRef]
41. Hallock, P. Changing influences between life and limestones in Earth history. In *Coral Reefs in the Anthropocene*; Birkeland, C., Ed.; Springer: Dordrecht, The Netherlands, 2015; pp. 17–42.
42. Mentel, M.; Martin, W. Energy metabolism among eukaryotic anaerobes in light of Proterozoic ocean chemistry. *Phil. Trans. R. Soc. B-Biol. Sci.* **2008**, *363*, 2717–2729. [CrossRef]
43. Ward, L.M.; Rasmussen, B.; Fischer, W.W. Primary productivity was limited by electron donors prior to the advent of oxygenic photosynthesis. *J. Geophys. Res. Biogeosci.* **2019**, *124*, 211–226. [CrossRef]
44. Fischer, W.W.; Hemp, J.; Johnson, J.E. Evolution of oxygenic photosynthesis. *Ann. Rev. Earth Planet. Sci.* **2016**, *44*, 647–683. [CrossRef]
45. Erb, T.J.; Zarzycki, J. A short history of RubisCO: The rise and fall (?) of Nature's predominant CO_2 fixing enzyme. *Curr. Opin. Biotech.* **2018**, *49*, 100–107. [CrossRef] [PubMed]
46. Badger, M.R.; Price, G.D. CO_2 concentrating mechanisms in cyanobacteria: Molecular components, their diversity and evolution. *J. Exper. Bot.* **2003**, *54*, 609–622. [CrossRef] [PubMed]
47. Raven, J.A.; Beardall, J.; Sánchez-Baracaldo, P. The possible evolution and future of CO_2-concentrating mechanisms. *J. Exper. Bot.* **2017**, *68*, 3701–3716. [CrossRef] [PubMed]
48. Papineau, D. Global biogeochemical changes at both ends of the Proterozoic: Insights from phosphorites. *Astrobiology* **2010**, *10*, 165–181. [CrossRef]
49. Large, R.R.; Halpin, J.A.; Lounejeva, E.; Danyushevsky, L.V.; Maslennikov, V.V.; Gregory, D.; Sack, P.J.; Haines, P.W.; Long, J.A.; Makoundi, C.; et al. Cycles of nutrient trace elements in the Phanerozoic ocean. *Gondwana Res.* **2015**, *28*, 1282–1293. [CrossRef]
50. Pufahl, P.K.; Groat, L.A. Sedimentary and igneous phosphate deposits: Formation and exploration: An invited paper. *Econ. Geol.* **2017**, *112*, 483–516. [CrossRef]
51. Paytan, A.; McLaughlin, K. The oceanic phosphorus cycle. *Chem. Rev.* **2007**, *107*, 563–576. [CrossRef]
52. Poulton, S.W.; Canfield, D.E. Ferruginous conditions: A dominant feature of the ocean through Earth's history. *Elements* **2011**, *7*, 107–112. [CrossRef]
53. Shen, J.; Webb, G.A.; Jell, J.S. Platform margins, reef facies and microbial carbonates; a comparison of Devonian reef complexes in the Canning Basin, Western Australia and the Guilin region, South China. *Earth-Sci. Rev.* **2008**, *88*, 33–59. [CrossRef]
54. Baturin, G.N. Phosphorus cycle in the ocean. *Lithol. Min. Res.* **2003**, *38*, 101–119. [CrossRef]
55. Chernoff, C.B.; Orris, G.J. Data set of world phosphate mines, deposits, and occurrences: Part A. geologic data; Part B. location and mineral economic data. *U.S. Geol. Surv. Open-File Rep.* **2002**, *C6*, 680.
56. Vescei, A.; Berger, W.H. Increase of atmospheric CO_2 during deglaciation: Constraints on the coral reef hypothesis from patterns of deposition. *Glob. Biogeochem. Cycles* **2004**, *18*, GB1035.
57. Danen-Louwerse, H.J.; Lijklema, L.; Coenraats, M. Coprecipitation of phosphate with calcium carbonate in Lake Veluwe. *Water Resour.* **1995**, *29*, 1781–1785. [CrossRef]
58. Gulbrandsen, R.A.; Cremer, M. Coprecipitation of carbonate and phosphate from seawater. *U.S. Geol. Surv. Prof. Pap.* **1970**, *700-C*, C125–C126.
59. Kitano, Y.; Okumura, M.; Idogaki, M. Uptake of phosphate ions by calcium carbonate. *Geochem. J.* **1978**, *12*, 29–37. [CrossRef]
60. Millero, F.; Huang, F.; Zhu, X.; Liu, X.; Zhang, J.-Z. Adsorption and desorption of phosphate on calcite and aragonite in seawater. *Aquat. Geochem.* **2001**, *7*, 33–56. [CrossRef]

61. Dupraz, C.; Reid, R.P.; Braissant, O.; Decho, A.W.; Norman, R.S.; Visscher, P.T. Processes of carbonate precipitation in modern microbial mats. *Earth-Sci. Rev.* **2009**, *96*, 141–162. [CrossRef]
62. Stal, L.J. Coastal microbial mats: The physiology of a small-scale ecosystem. *S. Afr. J. Bot.* **2001**, *67*, 399–410. [CrossRef]
63. Van Mooy, B.A.S.; Rocap, G.; Fredricks, H.F.; Evans, C.T.; Devol, A.H. Sulfolipids dramatically decrease phosphorus demand by picocyanobacteria in oligotrophic marine environments. *Proc. Natl. Acad. Sci. USA* **2006**, *103*, 8607–8612. [CrossRef]
64. McConnaughey, T.A.; Whelan, J.F. Calcification generates protons for nutrient and bicarbonate uptake. *Earth-Sci. Rev.* **2006**, *42*, 95–117. [CrossRef]
65. Visscher, P.T.; Hoeft, S.E.; Surgeon, T.-M.L.; Rogers, D.R.; Bebout, B.M.; Thompson, J.S.; Reid, R.P. Microelectrode measurements in stromatolites. Unraveling the Earth's past? In *Environmental Electrochemistry: Analyses of Trace Element Biogeochemistry*; ACS Symposium Series 811; Taillefert, M., Rozan, T., Eds.; Oxford University Press: New York, NY, 2002; pp. 265–282.
66. Wright, D.T.; Altermann, W. Microfacies development in Late Archaean stromatolites and oolites of the Ghaap Group of South Africa. *Geol. Soc. Lond. Spec. Pub.* **2000**, *178*, 51–70. [CrossRef]
67. Batchelor, M.T.; Burne, R.V.; Henry, B.I.; Li, F.; Paul, P. A biofilm and organomineralisation model for the growth and limiting size of ooids. *Sci. Rep.* **2018**, *8*, 559. [CrossRef] [PubMed]
68. Grotzinger, J.P.; James, N.P. Precambrian carbonates: Evolution of understanding. In *Carbonate Sedimentation and Diagenesis in the Evolving Precambrian World*; Grotzinger, J.P., James, N.P., Eds.; SEPM Society for Sedimentary Geology Special Publications: Tulsa, OK, USA, 2000; Volume 67, pp. 3–20.
69. Brehm, U.; Krumbein, W.E.; Palinska, K.A. Biomicrospheres generate ooids in the laboratory. *Geomicrobiol. J.* **2006**, *23*, 545–550. [CrossRef]
70. Diaz, M.R.; Eberli, G.P.; Blackwelder, P.; Phillips, B.; Swart, P.K. Microbially mediated organomineralization in the formation of ooids. *Geology* **2017**, *45*, 771–774. [CrossRef]
71. Diaz, M.R.; Eberli, G.P. Decoding the mechanism of formation in marine ooids: A review. *Earth-Sci. Rev.* **2019**, *190*, 536–556. [CrossRef]
72. Yates, K.K.; Robbins, L.L. Microbial lime-mud production and its relation to climate change. In *Geological Perspectives of Global Climate Change*; Gerhard, L.C., Harrison, W.H., Hanson, B.M., Eds.; American Association of Petroleum Geologists: Tulsa, OK, USA, 2001; Volume 47, pp. 267–283.
73. Robbins, L.L.; Tao, Y.; Evans, C.A. Temporal and spatial distribution of whitings on Great Bahama Bank and a new lime mud budget. *Geology* **1997**, *25*, 947–950. [CrossRef]
74. Robbins, L.L.; Blackwelder, P.L. Biochemical and ultrastructural evidence for the origin of whitings: A biologically induced calcium carbonate precipitation mechanism. *Geology* **1992**, *20*, 464–468. [CrossRef]
75. Bhattacharya, D.; Medlin, L. The phylogeny of plastids: A review based on comparisons of small-subunit ribosomal RNA coding regions. *J. Phycol.* **1995**, *31*, 489–498. [CrossRef]
76. Stiller, J.W.; Hall, B.D. The origin of the red algae: Implications for plastid evolution. *P. Natl. Acad. Sci. USA* **1997**, *94*, 4520–4525. [CrossRef]
77. Demes, K.W.; Littler, M.M.; Littler, D.S. Comparative phosphate acquisition in giant-celled rhizophytic algae (Bryopsidales, Chlorophyta): Fleshy vs. calcified forms. *Aquat. Bot.* **2010**, *92*, 157–160. [CrossRef]
78. Wizemann, A.; Meyer, F.W.; Westphal, H. A new model for the calcification of the green macro-alga *Halimeda opuntia* (*Lamouroux*). *Coral Reefs* **2015**, *33*, 951–964. [CrossRef]
79. Wizemann, A.; Meyer, F.W.; Hofmann, L.C.; Wild, C.; Westphal, H. Ocean acidification alters the calcareous microstructure of the green macro-alga *Halimeda opuntia*. *Coral Reefs* **2015**, *34*, 941–954. [CrossRef]
80. Nash, M.C.; Diaz-Pulido, G.; Harvey, A.S.; Adey, W. Coralline algal calcification: A morphological and process-based understanding. *PLoS ONE* **2019**, *14*, e0221396. [CrossRef] [PubMed]
81. Comeau, S.; Edmunds, P.J.; Spindel, N.B.; Carpenter, R.C. The responses of eight coral reef calcifiers to increasing partial pressure of CO_2 do not exhibit a tipping point. *Limnol. Oceanogr.* **2003**, *58*, 388–398. [CrossRef]
82. Hurd, C.L.; Cornwall, C.E.; Currie, K.; Hepburn, C.D.; McGraw, C.M.; Hunter, K.A.; Boyd, P.W. Metabolically induced pH fluctuations by some coastal calcifiers exceed projected 22nd century ocean acidification: A mechanism for differential susceptibility? *Glob. Chang. Biol.* **2011**, *17*, 3254–3262. [CrossRef]
83. Hay, W.W. Carbonate fluxes and calcareous nannoplankton. In *Coccolithophores–From Molecular Processes to Global Impact*; Thierstein, H., Young, J.R., Eds.; Springer: New York, NY, USA, 2004; pp. 509–527.
84. Monteiro, F.M.; Bach, L.T.; Brownlee, C.; Bown, P.; Rickaby RE, M.; Poulton, A.J.; Tyrrell, T.; Beaufort, L.; Dutkiewicz, S.; Gibbs, S.; et al. Why marine phytoplankton calcify. *Sci. Adv.* **2016**, *2*, e1501822. [CrossRef]
85. Young, J.R. Possible functional interpretations of coccolith morphology. *Abh. Geolog. Bundesanst.-A* **1987**, *39*, 305–313.
86. Paasche, E. Roles of nitrogen and phosphorus in coccolith formation in *Emiliania huxleyi* (Prymnesiophyceae). *Eur. J. Phycol.* **1998**, *33*, 33–42. [CrossRef]
87. Sheward, R.M.; Poulton, A.J.; Gibbs, S.J.; Daniels, C.J.; Bown, P.R. Physiology regulates the relationship between coccosphere geometry and growth phase in coccolithophores. *Biogeosciences* **2017**, *14*, 1493–1509. [CrossRef]
88. Brownlee, C.; Wheeler, G.L.; Taylor, A.R. Coccolithophore biomineralization: New questions, new answers. *Sem. Cell Devel. Biol.* **2015**, *46*, 11–16. [CrossRef]

89. Walker, C.E.; Heath, S.; Salmon, D.L.; Smirnoff, N.; Langer, G.; Taylor, A.R.; Brownlee, C.; Wheeler, G.L. An extracellular polysaccharide-rich organic layer contributes to organization of the coccosphere in coccolithophores. *Front. Mar. Sci.* **2018**, *5*, 306. [CrossRef]
90. Sviben, S.; Gal, A.; Hood, M.A.; Bertinetti, L.; Politi, Y.; Bennet, M.; Krishnamoorthy, P.; Schertel, A.; Wirth, R.; Sorrentino, A.; et al. A vacuole-like compartment concentrates a disordered calcium phase in a key coccolithophorid alga. *Nat. Com.* **2016**, *7*, 11228. [CrossRef] [PubMed]
91. Gal, A.; Sviben, S.; Wirth, R.; Schreiber, A.; Lassalle-Kaiser, B.; Faivre, D.; Scheffel, A. Trace-element incorporation into intracellular pools uncovers calcium-pathways in a coccolithophore. *Adv. Sci.* **2017**, *4*, 1700088. [CrossRef] [PubMed]
92. Pomar, L.; Hallock, P. Carbonate factories: A conundrum in sedimentary geology. *Earth-Sci. Rev.* **2008**, *87*, 134–169. [CrossRef]
93. Newell, N.D. Mass extinctions at the end of the Cretaceous Period. *Science* **1965**, *149*, 922–924. [CrossRef]
94. Keller, G.; Mateo, P.; Punekar, J.; Khozyem, H.; Gertsch, B.; Spangenberg, J.; Bitchong, A.M.; Adatte, T. Environmental Changes during the Cretaceous-Paleogene Mass Extinction and Paleocene-Eocene Thermal Maximum: Implications for the Anthropocene. *Gondwana Res.* **2018**, *56*, 69–89. [CrossRef]
95. Hallock, P. Fluctuations in the trophic resource continuum: A factor in global diversity cycles? *Paleoceanography* **1987**, *2*, 457–471. [CrossRef]
96. d'Halloy, J.-J. Observations sur un essai de carte géologique de la France, des Pays-Bas, et des contrées voisines [*Observations on a trial geological map of France, the Low Countries, and neighboring countries*]. *Ann. Des Mines* **1822**, *7*, 353–376.
97. Lear, C.H.; Rosenthal, Y.; Wright, J.D. The closing of a seaway: Ocean water masses and global climate change. *Earth Planet. Sci. Lett.* **2003**, *210*, 425–436. [CrossRef]
98. Adams, C.G.; Lee, D.E.; Rosen, B.R. Conflicting isotopic and biotic evidence for tropical sea-surface temperatures during the Tertiary. *Palaeogeog. Palaeoecol.* **1990**, *77*, 289–313. [CrossRef]
99. Stanley, S.M.; Hardie, L.A. Secular oscillations in the carbonate mineralogy of reef-building and sediment-producing organisms driven by tectonically forced shifts in seawater chemistry. *Palaeogeog. Palaeoecol.* **1998**, *144*, 3–19. [CrossRef]
100. Pearson, P.N.; Palmer, M.R. Atmospheric carbon dioxide concentrations over the past 60 million years. *Nature* **2000**, *406*, 695–699. [CrossRef] [PubMed]
101. Pochon, X.; Montoya-Burgos, J.I.; Stadelmann, B.; Pawlowski, J. Molecular phylogeny, evolutionary rates, and divergence timing of the symbiotic dinoflagellate genus *Symbiodinium*. *Mol. Phylogenet. Evol.* **2006**, *38*, 20–30. [CrossRef] [PubMed]
102. Wooldridge, S.A. Breakdown of the coral-algae symbiosis: Towards formalising a linkage between warm-water bleaching thresholds and the growth rate of the intracellular zooxanthellae. *Biogeosciences* **2013**, *10*, 1647–1658. [CrossRef]

MDPI
St. Alban-Anlage 66
4052 Basel
Switzerland
Tel. +41 61 683 77 34
Fax +41 61 302 89 18
www.mdpi.com

Journal of Marine Science and Engineering Editorial Office
E-mail: jmse@mdpi.com
www.mdpi.com/journal/jmse